Springer Proceedings in Business and Economics

Springer Proceedings in Business and Economics brings the most current research presented at conferences and workshops to a global readership. The series features volumes (in electronic and print formats) of selected contributions from conferences in all areas of economics, business, management, and finance. In addition to an overall evaluation by the publisher of the topical interest, scientific quality, and timeliness of each volume, each contribution is refereed to standards comparable to those of leading journals, resulting in authoritative contributions to the respective fields. Springer's production and distribution infrastructure ensures rapid publication and wide circulation of the latest developments in the most compelling and promising areas of research today.

The editorial development of volumes may be managed using Springer Nature's innovative EquinOCS, a proven online conference proceedings submission, management and review system. This system is designed to ensure an efficient timeline for your publication, making Springer Proceedings in Business and Economics the premier series to publish your workshop or conference volume.

This book series is indexed in SCOPUS.

Anna K. Zarkada
Editor

Empowering Digital Transformation

Proceedings of the 2025 GMA-GAMMA Joint Symposium, Hong Kong, July 24–27, 2025

Springer

Editor
Anna K. Zarkada
Department of Communication and Marketing
Cyprus University of Technology
Limassol, Cyprus

ISSN 2198-7246 ISSN 2198-7254 (electronic)
Springer Proceedings in Business and Economics
ISBN 978-3-031-96929-4 ISBN 978-3-031-96930-0 (eBook)
https://doi.org/10.1007/978-3-031-96930-0

This Springer imprint is published by the registered company Springer Nature Switzerland AG
The registered company address is: Gewerbestrasse 11, 6330 Cham, Switzerland

Preface

Following the success of the GMA-GAMMA Joint Symposia of 2020 and 2023 and the publication of the 2023 proceedings in a book entitled Marketing Solutions to the Challenges of a VUCA Environment in the Springer Proceedings in Business and Economics Series, the Greek Marketing Academy (GMA) and the Global Alliance of Marketing & Management Associations (GAMMA) hosted another successful Joint Symposium in the 2025 Global Marketing Conference (GMC) at Hong Kong.

Selected Symposium papers were invited to be expanded and further elaborated, and together with other cutting edge research contributions, they form the chapters of this book. The book explores the theme of Empowering Digital Transformation going beyond simply documenting the process and effects of digital transformation to explore what marketing and management can do to engage and empower people whilst efficiently integrating processes and platforms.

Anna K. Zarkada

Contents

The Digital Turn: Reimagining Business, Marketing, and the Consumer in a Techno-Driven Age

Panagiotis Moiras(✉)

Cyprus University of Technology, 3036 Limassol, Cyprus
pk.moiras@edu.cut.ac.cy

1 Introduction

The digital transformation of the twenty-first century is not merely a technological shift—it is a profound socio-economic reconfiguration, redefining how businesses operate, industries compete, and consumers engage. As Klaus (2016), founder of the World Economic Forum, stated, we are *"at the beginning of a revolution that is fundamentally changing the way we live, work and relate to one another"*. Artificial Intelligence (AI), big data, blockchain, and automation now shape modern commerce, fostering innovation, transparency, and hyper-connectivity.

In business and marketing, AI-driven consumer profiling, blockchain-enabled supply chains, and AR-based product testing enhance eficiency, targeting, and engagement. Tapscott and Williams (2008) describe this shift as "wikinomics," where decentralization and collaboration replace rigid hierarchies. Companies now focus on experiences rather than just products. Sephora's virtual try-on apps and Nike's online shoe customization are notable examples of consumers empowerment to co-create value. Such examples highlight the transformative potential digitalisation holds for reshaping customer interactions and redefining value propositions.

The significant benefits derived from digital transition have not gone unnoticed in academic literature, which increasingly examines its strategic and operational advantages. Artificial Intelligence optimises operations by automating tasks, allowing human resources to concentrate on strategic initiatives (AlNuaimi et al., 2022). Huang and Rust (2021) emphasise the significance of AI in marketing, encompassing automation and comprehensive emotional analysis, hence enhancing segmentation and targeting. Omnichannel marketing and behavioural targeting are crucial, as big data facilitates personalised engagement (Mavlutova et al., 2022). Moreover, according to McKinsey Global Institute (2018), digitally advanced companies exceed their competitors' performance by 23%. Digital channels have revolutionised marketing, enhancing transparency, trust, and economic development, especially for SMEs (Jeza & Lekhanya, 2022).

However, critics question whether digitalization truly empowers or fosters new power imbalances. Zubof (2019) warns of "surveillance capitalism," where personal data is exploited without consent, while at the same time, Morozov (2013) critiques "technological solutionism," reducing societal issues to mere coding problems. Digital inequality

A. K. Zarkada (Ed.): GMA-GAMMA 2025, SPBE, pp. 1–8, 2025.
https://doi.org/10.1007/978-3-031-96930-0_1

also remains a concern. SMEs, especially in emerging economies, struggle with limited resources and infrastructure, deepening the digital divide. Algorithmic bias further exacerbates exclusion, as seen in Amazon's flawed AI hiring tool (Dastin, 2022).

Beyond operational and ethical concerns, lies an existential question: what becomes of human agency in the age of automation? Philosopher Byung-Chul Han (2017) warns that hyper-eficiency risks eroding the contemplative slowness that nurtures meaning and connection.

We're shifting from a world where economic, social, and cultural activity is driven by profit, to one dominated by *rent*—not rent for physical land but rent for access to digital territory: the cloud. Like the land enclosures of the 18th century that enabled commodification, today's digital enclosures have privatized the Internet. Big Tech, fused with finance and private equity, has created a system that no longer resembles capitalism—because its two pillars, markets and profits, have been dismantled. This is an era that economist and former Greek Minister of Finance Yanis Varoufakis calls *technofeudalism*, which is not characterised just by market concentration; it's the end of the market itself (Varoufakis, 2023). Interaction with Amazon is a vivid manifestation of this notion. When Amazon recommends a product, it's uncannily accurate. It doesn't just predict what I want—it shapes my preferences over time. Through interactions with tools like Alexa, a feedback loop is formed: I influence it, it influences me, and eventually, it sells to me directly, bypassing the traditional market altogether. Clearly, such a reality departs from Adam Smith's classical vision of the marketplace, characterized by free competition, self-regulation, and the guiding force of the "invisible hand" (Smith, 1793).

Therefore, digital transformation is not simply a technological imperative but a philosophical and strategic one. The challenge is not just in digitising operations but in humanising them. Organisations must go beyond tech adoption and engage in ethical introspection, designing systems that are inclusive, transparent, and accountable.

2 Thematic Overview of the Chapters

The digital transformation of businesses is no longer a distant concept but a present-day reality. Organisations, both large and small, must not only adapt to this new landscape but also develop strategic frameworks that ensure their long-term sustainability in an increasingly digital world. This book, *Empowering Digital Transformation*, explores how digitalisation influences corporate strategies, consumer behaviours, marketing practices, and industry-specific applications. The chapters are structured into those four key thematic sections, each offering a unique perspective on how digital evolution is reshaping the business and marketing ecosystem.

3 Foundations of Digital Transformation in Business and Marketing

The first section of this book sets the groundwork for understanding the strategic implications of digital transformation in modern business practices. It explores how organisations can integrate emerging technologies into their operations while maintaining a balance between innovation, corporate sustainability, and human-centric approaches.

Marinos Voukis, in his chapter *Exploring the Intersection of Digital Evolution and Corporate Sustainability and Responsibility* (Ch.1), argues that digital transformation extends far beyond technological adoption—it is deeply intertwined with corporate sustainability and responsibility (CSR). Contemporary enterprises must adopt new digital tools while ensuring these technologies promote transparency, eficiency, and involvement within sustainable business goals. This chapter illustrates how artificial intelligence (AI), the Internet of Things (IoT), and blockchain might promote more accountable corporate practices. Nonetheless, as the author emphasises, incorporating digital technologies into CSR projects presents problems, requiring organisations to address ethical dilemmas and implementation obstacles to attain significant advancement.

A similar perspective is echoed in *Paving the Way for Digital Transformation in Contemporary Organizations Through Building Strong Organizational Capabilities* (Ch.2), by Sofia Chatzi, which highlights the necessity of aligning human resources with technological advancements. Digital transformation is frequently perceived as the implementation of software, automation, and data-informed decision-making. This chapter contests that idea by offering a comprehensive viewpoint that emphasises the significance of leadership, workforce flexibility, and organisational agility in attaining successful digitisation. The report elucidates that strategic human resource alignment and robust leadership are crucial for navigating people through the intricacies of transition, guaranteeing that digitisation transcends a superficial enhancement to become an integral organisational shift.

Nicos Rossides further examines the human aspect of digital transformation in *Bridging Bytes and Behaviors: Human-Centric Research in Emerging Digital Economies* (Ch.3). Technological innovation provides organisations with robust tools for data-driven decision-making, although it also prompts significant enquiries over consumer behaviour, cultural adaption, and privacy. This chapter examines the utilisation of AI-driven analytics in international marketing research, specifically within growing digital economies like the Gulf States, China, and India. The study demonstrates that technical solutions must be tailored to various cultural settings to efectively adapt digital ecosystems to local requirements. The chapter contends that although big data and automation yield significant insights, they must be contextualised within human-centric frameworks to guarantee ethical and efective use.

4 Digitalization of Consumer and Buyer-Seller Interactions

The digital revolution of consumer behaviour is a significant change influencing contemporary marketing techniques. The advent of social media, AI-powered analytics, and virtual engagement technologies has transformed business-customer interactions. This section examines the impact of digitalisation on buyer-seller interactions, consumer expectations, and data-driven marketing techniques.

The philosophical implications of digitalisation are explored in *The Evolution of Buyer- Seller Relationships in the Digital Age: A Philosophical Approach* (Ch.4) by Erasmia Leonidou. As commerce moves further into the digital sphere, fundamental questions arise about trust, authenticity, and ethical responsibility in consumer-business interactions. This chapter analyses the impact of digitalisation on buyer-seller relationships, interrogating the extent to which online transactions can authentically replicate the personal, trust-centric interactions that have historically characterised commerce. It critically explores how digital commerce has simultaneously enriched and complicated these dynamics, necessitating a re-evaluation of customer relationship strategies in an increasingly automated environment.

Another aspect of digital transformation is the rise of social media as a dominant force in consumer research. Yioula Melanthiou, in *The Social Shift: How Social Media Revolutionised Consumer Behaviour Research* (Ch.5) presents a compelling case for how social media platforms have become indispensable tools for understanding purchasing behaviour. Conventional research methodologies, including surveys and focus groups, are progressively being supplanted by real-time, extensive digital data analysis. This transition has enabled marketers to analyse audience behaviour with unparalleled precision, monitoring trends, preferences, and interaction patterns. This chapter emphasises the dificulties of utilising social media for consumer research, particularly ethical issues of data privacy and the impact of digital algorithms on buying behaviour.

One of the most innovative applications of digital transformation in consumer interaction is examined in *The Impact of Virtual Try-On Tools for Beauty Products on Consumer Behavior: A Segmentation Study* (Ch.6) by Anna Zarkada and Yioula Melanthiou. The beauty industry has been at the forefront of adopting augmented reality (AR) to enhance consumer engagement, and this chapter explores how virtual try-on (VTO) tools are transforming the way customers shop for cosmetics. The segmentation analysis reveals two primary customer segments: *hedonic enthusiasts*, who relish the immersive nature of virtual trials, and *utility-focused* consumers, who prioritise eficiency and accuracy in their purchasing choices. The results offer strategic insights for beauty businesses aiming to customise their digital strategy according to varying consumer preferences.

5 AI, Automation, and Digital Transformation in Industry-Specific Applications

Although digital transformation is a worldwide occurrence, its effects differ among industries. This section examines the implementation of AI, automation, and digital marketing techniques in essential economic areas, specifically tourism and travel.

AI has become an essential component of the tourism industry, as examined in *Exploring the Impact of Artificial Intelligence on the Tourism Industry: Opportunities and Challenges* (Ch.7) by Ioannis Katsanakis, Gorgia Zouni, Stergios Spasos, Dimitra-Margieta Lykoudi and Nikolaos Akriotis. This chapter explores how AI-driven automation is reshaping the travel experience, from personalised customer service to predictive analytics for demand forecasting. The implementation of AI offers significant prospects for efficiency and cost savings, although it also raises obstacles including employment displacement and ethical issues related to data security. The chapter offers a balanced examination of AI's transformational capabilities while simultaneously warning of its unforeseen repercussions.

Complementing this discussion, *Information Systems and Marketing of Tour Operators: Impacts, User Perspectives, Challenges, and Future Directions* (Ch.8) by Anna Krousaraki and Ourania Vitouladiti, examines how digital marketing and information technologies have transformed the travel and tourism industry. The emergence of online booking platforms and AI-driven recommendation systems has granted tour operators unparalleled access to customer data. This chapter analyses the ways in which digital tools strengthen marketing initiatives, increase customer relationship management, and optimise operations, while simultaneously confronting the challenges posed by data saturation and consumer trust.

6 Digital Advertising and Marketing Strategies for SMEs

Small and medium-sized organisations (SMEs) have distinct obstacles in the pursuit of digital transformation. Although digital tools enable SMEs to compete with larger firms, they frequently face obstacles including budgetary limitations, regulatory compliance, and technical intricacies. This section examines how SMEs can leverage digital advertising and marketing technologies to enhance their competitive advantage.

In *Benefits and Perils of SMEs' Digital Transformation in Marketing* (Ch.9) by Meletios Niros, Eirini Peitzika, Angelica Niros and Panagiota Kelali, the focus is on how SMEs can leverage digital tools to enhance marketing eficiency. The chapter presents case studies of SMEs that have successfully integrated AI, big data analytics, and programmatic advertising into their strategies. However, the study also highlights consumer scepticism towards digital marketing tactics, as well as the security risks and financial hurdles that SMEs must overcome.

The final chapter (Ch.10), SMEs Playing in the Big Advertising League: The Transformative Potential of Programmatic Advertising and the Barriers to Its Full-Scale Implementation by Marius Ian Yul Horstick and Anna Zarkada, delves into the opportunities and challenges of programmatic advertising. This technology enables SMEs to accurately target consumers and optimise their advertising expenditure; yet, adoption is constrained by regulatory concerns and implementation complexities. The chapter presents a four-phase framework to assist SMEs in overcoming these hurdles, offering practical ways for achieving success in digital marketing.

7 Synopsis of Key Reflections

Although the book provides a solid framework for companies starting their digital transformation journey, reading it with an analytical eye shows that digitalisation is really about more than just using technology. In digital transformation, technology and humanity are in a constant state of conversation, with each side influencing and changing the other. The goal of digital transformation should be to bring about significant social development and genuine human relationships; yet, if progress is made at the expense of direction and ethical consciousness, we risk losing sight of this essential.

In their critical reflections on the ways digitalisation, corporate sustainability, and human-centricity interact with one another, the chapters reveal an unstated but necessary synergy. Tech solutions do not automatically result in ethical business practices, despite the fact that many companies are praising digitalisation for its ability to improve sustainability initiatives via blockchain, artificial intelligence, or the internet of things (IoT). An even more perceptive analysis highlights the urgent need for a paradigm change, one in which sustainability is seen as intrinsic to corporate goals rather than an afterthought. Consequently, genuine digital sustainability requires deeper integration of digital ethics, transparency, and corporate accountability, echoing Microsoft CEO Satya Nadella's insight: *"Our industry does not respect tradition—it only respects innovation. But innovation must have purpose, and technology must empower humanity rather than diminish it"* (Nadella et al., 2018).

To enhance our comprehension, it is beneficial to explore the practical ramifications of integrating digital sustainability into organisational frameworks. In this regard, one should contemplate these pivotal insights that elucidate the fundamental relationship among ethics, culture, and the adoption of purpose-driven technology:

- **Digital Ethics Integration:** Organisations should proactively embed ethical considerations into technological innovations rather than retroactively addressing them.
- **Cultural Agility:** Digital transformation demands not just technological agility but also a profound shift in organisational culture toward ongoing learning and adaptability.
- **Technology with Purpose:** Each technological adoption should be strategically evaluated based on its ability to enhance human well-being, corporate responsibility, and societal value—not merely operational efficiency.

Furthermore, contemplation of the decline of trust in digital settings is prompted by a philosophical examination of buyer-seller interactions in the digital era. The advent of digital technologies has altered the dynamics of consumer spending and has also called into question the foundations of credibility and reliability. Virtual try-ons and AI analytics can help businesses complete sales, but meaningful connections need digital interactions that go beyond just making a sale to establish trust.

Organisations must critically reassess their digital strategies to ensure they remain grounded in human values. Additional considerations include:

- **Reimagining Trust Digitally:** Businesses must find innovative ways to replicate personal trust digitally—perhaps through greater transparency, consistent authenticity, and accountability mechanisms.
- **Human-Technology Balance:** Leveraging data analytics and AI must not overshadow the critical role of empathy, intuition, and human judgment in customer interactions.

Ultimately, digital transformation in SMEs underscores a crucial yet sometimes disregarded lesson: agility alone is inadequate. Small and medium-sized enterprises must foster technological proficiency, ethical awareness, and flexibility. Adopting digital advertising techniques such as programmatic advertising requires SMEs to comprehend both their tools and the significant ethical ramifications with data privacy and targeted marketing. To prosper in the face of digital disruption, SMEs must attain digital acumen, understanding that genuine empowerment stems from a strategic equilibrium between technology innovation and human-centered ethical responsibility. As former IBM CEO Ginni Rometty emphasises: *"Data is the phenomenon of our time; it is the world's new natural resource. But the value isn't in data itself—it's in what you do with it responsibly"* (Rometty, 2018).

Additional practical takeaways for SMEs include:

- **Digital Literacy & Responsibility:** SMEs should prioritise continuous education about emerging digital technologies and related ethical standards.
- **Strategic Alignment:** Digital marketing strategies must align closely with authentic brand values to maintain consumer trust.
- **Balanced Innovation:** SMEs should innovate sustainably, weighing technological benefits against ethical responsibilities and data privacy concerns.

This synthesis of those insights urges organisations to move beyond mere digital adoption and foster a more responsible, sustainable, and human-centric digital future.

8 Conclusion

The chapters in this book collectively offer a thorough overview of the shifting landscape of digital transformation. Each piece of research, provides vital insights into core tactics, shifts in consumer behaviour, industry-specific applications, and marketing solutions for SMEs, enabling firms to adeptly navigate this era of swift digital transformation. As digital revolution reshapes commerce, companies must stay adaptable, morally aware, and strategically equipped to seize the forthcoming opportunities and challenges.

Acknowledgements/List of Reviewers. We gratefully acknowledge the commitment and diligence of the reviewers, whose thoughtful comments and valuable suggestions substantially improved the academic rigor, clarity, and relevance of each chapter. Their meticulous efforts have been essential in elevating the overall quality of this volume.

Apostolidis Chrysostomos
Kashif Muhammad
Katsanakis Ioannis
Koronaki Eirini
Koundouri Phoebe
Kyriacou Evdoxia
Leonidou Costas
Leonidou Erasmia
Melanthiou Yioula
Niros Meletios
Panopoulos Anastasios
Papageorgiou Athina
Papasolomou Ioanna
Peitzika Eirini
Rossides Nicos
Rossides Nicos
Sarantakou Efthymia
Stergiou Dimitrios
Tsaousi Christiana
Tzoumaka Eugenia
Vitouladiti Ourania
Voukis Marinos
Voutsa Maria
Zarkada Anna.
Zouni Georgia

References

AlNuaimi, B.K., Kumar Singh, S., Ren, S., Budhwar, P., Vorobyev, D.: Mastering digital transformation: the nexus between leadership, agility, and digital strategy. J. Bus. Res. **145**, 636–648 (2022). https://doi.org/10.1016/j.jbusres.2022.03.038

Dastin, J.: Amazon scraps secret AI recruiting tool that showed bias against women. In: Ethics of Data and Analytics, pp. 296–299. Auerbach Publications (2022)

Han, B.-C.: In the Swarm: Digital Prospects (Vol. 3). MIT press, Cambridge (2017)

Huang, M.-H., Rust, R.T.: A strategic framework for artificial intelligence in marketing. J. Acad. Mark. Sci. **49**, 30–50 (2021)

Institute, M. G. Notes from the AI frontier: Applications and value of deep learning (2018). https://www.mckinsey.com/featured-insights/artificial-intelligence/notes-from-the-ai-frontier-applications-and-value-of-deep-learning

Jeza, S., Lekhanya, L.M.: The influence of digital transformation on the growth of small and medium enterprises in South Africa. Probl. Perspect. Manag. **20**(3), 297 (2022)

Klaus, S.: The fourth industrial revolution. In: Moscow: Ekonomika (2016)

Mavlutova, I., Spilbergs, A., Verdenhofs, A., Natrins, A., Arefjevs, I., Volkova, T.: Digital transformation as a driver of the financial sector sustainable development: an impact on financial inclusion and operational eficiency. Sustainability **15**(1), 207 (2022)

Morozov, E.: To save everything, click here: The folly of technological solutionism. PublicAfairs (2013)

Nadella, S., Shaw, G., Nichols, J.T.: Hit Refresh Intl: The Quest to Rediscover Microsoft's Soul and Imagine a Better Future for Everyone. In: Harper Business (2018)

Rometty, G.: Keynote Speech, World Economic Forum Annual Meeting, Davos 2018 (2018). In

Smith, A.: An Inquiry Into the Nature and Causes of the Wealth of Nations: By Adam Smith (Vol. 1). William Porter (1793)

Tapscott, D., Williams, A.D.: Wikinomics: how mass collaboration changes everything. Penguin (2008)

Varoufakis, Y.: Capitalism is dead: long live Technofeudalism - We have all been turned into cloud-serfs (2023). https://unherd.com/2023/09/capitalism-is-dead-long-live-technofeudalism/

Zubof, S.: Surveillance Capitalism and the Challenge of Collective Action. In New labor forum (Vol. 28, No. 1, pp. 10–29). Sage CA Publications, Los Angeles, CA (2019)

Foundations of Digital Transformation in Business and Marketing

Exploring the Intersection of Digital Evolution and Corporate Sustainability and Responsibility

Marinos Voukis[1,2](✉)

[1] Director of Communication, Marketing and International Relations Services, Cyprus University of Technology, P.O. Box. 50329, 3603 Limassol, Cyprus
marinos.voukis@cut.ac.cy

[2] Chairman, Board of Directors, CSR Cyprus 4, Cyprus Employers and Industrialists Federation Street, 2000 Strovolos, Nicosia, Cyprus

Abstract. This research delves into the dimensions of digital evolution highlighting the importance of fostering a culture that values innovation, adaptability and continual progress. Additionally, it delves into Corporate Sustainability and Responsibility (CSR) concepts and explores how they intersect with evolution. The above aim to explore the intersection of digital evolution and CSR, examining how emerging digital technologies impact sustainable corporate strategies. The study emphasizes the challenges and opportunities that organizations face when they embed digital technology into their sustainability initiatives. This research is based on: a) a systematic literature review and b) a conceptual analysis of recent academic and industry sources. It combines vital conclusions from academic articles, case studies and reports, in order to provide a structured perspective on the growing relationship between Digital Transformation (DT) and CSR. The study identifies specific core areas in which DT is positively affecting CSR and it also highlights concerns and challenges that are associated with the incorporation of DT in CSR initiatives. It aims to contribute to the existing discussion of two important areas, the DT and the CSR and most importantly, to the rising discussion of the intersection of these two areas, by adding an investigation of the effect of DT on CSR.

Keywords: CSR · DT · DigitalTransformation · Sustainability

1 Introduction

Digital evolution involves adopting technologies to redefine business operations and enhance customer interactions. It requires a shift in mindset that challenges business models while promoting innovation and flexibility (Vial, 2021; Verhoef et al., 2021). In today's business landscape the role of CSR is increasingly vital as organizations aim to meet stakeholder expectations. Digital Transformation (DT) significantly affects an organization's sustainability aspects through operation, customer, business model, technology, workforce, and collaboration (Tewes et al. 2023; Schallmo et al. 2024; Nosratabadi et al. 2023). Digital evolution refers to the incorporation of technology,

A. K. Zarkada (Ed.): GMA-GAMMA 2025, SPBE, pp. 11–24, 2025.
https://doi.org/10.1007/978-3-031-96930-0_2

throughout an organization resulting in significant changes to how operations and value are delivered. Through the shift in mindset that challenges business models while promoting innovation and flexibility and the adaptation of new technologies, the business operations are refined and the customer interactions are enhanced (Qing & Jin, 2023).

The integration of transformation can greatly boost a company's CSR endeavors by offering the tools and capabilities to enhance transparency, efficiency and engagement. Key areas where DT aids CSR efforts include: a) Utilizing Data for Informed Decision Making and b) Enhancing Transparency and Reporting.

2 Integration of Technology

The integration of cutting-edge technologies (e.g. cloud computing, artificial intelligence (AI) machine learning, Internet of Things (IoT), blockchain) facilitate and advance the evolution as they not only transform processes but also products and services. These advancements help companies boost productivity expand scalability and foster creativity (Zhang et al. 2023; Zhang et al. 2022; World Economic Forum, 2024; Schallmo et al., 2017; Westerman et al., 2014).

Cloud computing offers computing resources that can be accessed as needed leading to cost savings and operational efficiencies. Businesses utilize cloud platforms to host applications, store data and support operations in an scalable manner (Armbrust et al., 2010).

AI and machine learning technologies are applied in various areas such, as customer service, predictive maintenance and tailored marketing strategies (Qing, C., & Jin, S. 2023; Russell & Norvig 2016). This results to the empowerment of organizations to analyze datasets recognize patterns and drive decisions based on data.

The IoT involves the linking of devices to the internet for collecting and analyzing data in real time. The positive outcomes derived vary from the boosting of efficiency and the enhancement of decision-making processes to, according to previous research (Ashton, 2009), the commencement of business possibilities in various sectors, e.g. manufacturing, healthcare and transportation.

Blockchain technology offers a secure method, for recording transactions and managing data. It improves transparency, decreases activities and can be utilized in fields, like managing supply chains, finance and healthcare (Tapscott & Tapscott 2016).

3 Enhancing Customer Experience

A goal of transformation is to enhance customer engagement. By using data analytics companies can provide customized multiplatform experiences. Predictive analytics help in understanding and predicting customer requirements and behaviors for a content customer base (Vogue Business, 2024; Lemon & Verhoef 2016; Wessel et al., 2021).

Personalization is requested by an increasing number of customers. Personalizing involves adjusting products, services and communications based on customer preferences. This can be accomplished through data analytics, AI and machine learning that analyze customer actions and choices to offer tailored experiences (Pine & Gilmore 1999; Yang et al. 2025).

In addition, omnichannel strategies merge touchpoints for customers to ensure a uniform experience across all platforms. This method guarantees that customers can engage with the brand through their favored channels while maintaining a positive experience (Verhoef et al., 2015; Li et al. 2023).

4 Optimizing Operational Processes and Innovation of Business Models

Simplifying and automating processes is crucial, for enhancing efficiency and cutting down expenses.

According to previous research (Wickremasinghe et al. 2024; Ross et al., 2016; Besson & Rowe 2012), the utilization of automation tools, the streamlining of supply chains and the fostering of communication between departments, are elements that drive this shift.

The automation technologies such as the robotic process automation (RPA) and the AI handle operational tasks. As a result, humans focus on activities that are targeting the organization's strategy which, according to other research (Ni et al. 2023; Aguirre et al., 2017; TIME 2024), it enhances efficiency and lowers the risk of errors.

The optimization of supply chains through solutions enables real time monitoring and management. IoT devices, technology and data analytics help in tracking inventory ensuring conditions and enhancing logistics for cost efficiency (Ivanov et al., 2019).

DT drives the innovation of business models by opening up market possibilities. This transition may involve shifting from product models to service oriented approaches building digital platforms or exploring fresh revenue streams through innovative digital channels (Qing & Jin 2023; Sund et al., 2020; Foss & Saebi 2017).

Embracing service-oriented business models means providing products as services, than one-time sales.

Businesses can offer services through subscriptions, leases or pay, per use models to provide value to customers and generate recurring revenue. Digital platforms bring together stakeholders like customers, suppliers and partners to engage and conduct transactions. These platforms help create ecosystems that innovation and add value.

5 Cultural and Leadership Dynamics

DT occurs provided that the existing culture embraces change, fosters innovation and promotes agility. The establishment of a culture that is open to change and receptive to experimentation entails the promotion of innovation. Positive attitude is essential in order to view failures as opportunities for learning. Agile organizations are characterized by their use of development processes, cross functional teams and a dedication, to improvement (Rigby et al., 2016; Westerman et al., 2014). Market trends should be constantly observed and evaluated in order to identify opportunities for new technologies that would serve the organization's strategic goals. Continuous improvement entails evaluating and enhancing processes, products and services. This fosters a culture of excellence and innovation (Liker, 2004).

For this reason, leadership plays a crucial role. Leadership must be constantly communicating a clear vision and strategy in order to ensure alignment and drive the DT process. Effective leadership is crucial, in steering DT efforts by defining a vision, establishing strategic objectives and motivating teams to embrace innovative work methods that are based on DT. To excel in this field one needs communication abilities, a grasp of technology, the skill to guide others through transitions and the ability to overcome challenges leaders face in aligning digital strategies with organizational goals (Buhse, 2023; Creusen, 2024; Moskaliuk & Ott 2023; Raja, 2023; Sharma, 2023; Stebbins, 2024; Westerman et al., 2014). According to the findings of a recent empirical study from 125 companies across Germany, Switzerland, France, and the UK to identify drivers of organizational agility while categorizing organizations into four profiles—laggards, execution specialists, experimenters, and leaders— and examining their approaches to agile transformation, higher agility levels are associated with a holistic approach, focusing on processes and products, with IT, product development, and research being the most agile functions. Additionally, agile organizations often employ multiple agile scaling frameworks (Kovynyov, et al., 2024).

6 Strategic Use of Data and Analytics

Data is seen as an asset, in today's landscape (Smith & Thompson, 2023a, b). Past research has highlighted the importance of data governance frameworks in maintaining data quality and ensuring compliance with regulatory standards (Nguyen & Lee 2024; Smith & Thompson, 2023a, b). According to the concept of data governance, the establishment of guidelines and protocols will ensure the precision, security and usability of data. Previous research (Abraham et al., 2019; Nguyen & Lee 2024), indicates that this encompasses managing data quality, adhering to privacy regulations and overseeing the lifecycle of data. Implementing data governance practices alongside data analytics facilitates well informed decision making and cultivates competitive edges.

According to previous research (Brynjolfsson & McAfee 2014; Davenport & Ronanki 2018; Patel & Singh, 2023a, b), predictive analytics and insights derived from data, play roles in shaping endeavors. Predictive analytics utilize data and machine learning algorithms in order to predict trends, events and behaviors. It finds applications across sectors like forecasting customer behavior predicting demands and managing risks, to aid decision making (Choudhury et al., 2019; Garcia & Lopez 2023a, b; Kumar & Sharma, 2025).

It should be highlighted that successful data-driven strategies adopted by organizations could lead to the achievement of a competitive advantage for the organizations as various analytical tools and frameworks help in understanding customer behavior, optimizing supply chains, and enhancing product development (Chen & Zhang, 2024a, b; Li & Wang, 2025; O'Connor & Murphy, 2023; Patel & Singh, 2023a, b; Smith & Nguyen, 2023a, b).

7 Security and Compliance Imperatives

When it comes to security and compliance imperatives robust cybersecurity measures play a role in safeguarding data integrity and upholding stakeholder trust, while also identifying potential risks, improving compliance, and enhancing financial performance. According to previous scholars, (Bowers & Connolly 2020; Madnick et al., 2009; Smith & Nguyen, 2023a, b), effective DT strategy requires compliance, with regulations and the implementation of security protocols.

Cybersecurity consists of measures that target the safeguarding of data, systems and networks from various cyber threats. According to other publication, in order to recognize and eliminate security holes effectively, the use of firewalls, encryption, multi-factor authentication and ongoing monitoring is essential (NIST, 2018). Other past research demonstrated how encryption technologies can be effectively utilized to maintain data integrity in cloud computing environments, while emphasizing the need for robust encryption standards (Patel & Lee 2024).

Regulatory compliance is an aspect for organizations requiring adherence to standards such as GDPR, HIPAA and ISO 27001 to safeguard data privacy. This involves setting policies conducting audits and providing staff with training on data protection practices (Garcia & Thompson, 2023; Zhang & Zhou 2018; O'Connor & Williams, 2025).

In terms of employee training and awareness, past research has illustrated the effectiveness of employee training programs in enhancing cybersecurity compliance, using case studies from various industries (Chen & Roberts, 2024).

Lastly, regular cybersecurity audits contribute to maintaining compliance with security standards in educational settings (Singh & Verma, 2023).

8 Corporate Sustainability and Responsibility (CSR)

Corporate Sustainability and Responsibility (CSR) refers to a company's dedication to the triple bottom line. It goes beyond profit making to ensure impacts on society and the environment. CSR covers activities, like business conduct, community involvement, environmental conservation and transparent governance (Carroll, 1999; Elkington; 1997). The concept of CSR has undergone changes, over the few decades. Initially centered around charity and compliance and named Corporate Social Responsibility, it has now expanded to encompass initiatives that're in line with a company's fundamental values and long-term business objectives (Carroll, 2008).

In today's business landscape the importance of CSR has grown as stakeholders—customers, employees, investors, regulators and more—expect levels of accountability and ethics. Planned CSR initiatives can boost a company's reputation build consumer confidence leading to term financial success (Financial Express 2024; Porter & Kramer; 2006).

8.1 Key Aspects

1. *Economic Responsibility;* Striving for profitability while upholding business standards. Generating wealth for shareholders while maintaining business practices, transparency and ethical behavior is at the core of responsibility. This involves management, fair trade pricing strategies and investments in innovation and sustainable development (Friedman, 1970). Additionally, it corresponds to actions for the reduction and/or elimination of poverty.
2. *Social Responsibility;* Contributing to community welfare. Enhancing quality of life. Social responsibility centers on how business activities affect society. It encompasses efforts to enhance employee well-being, support community growth and uphold rights. Additionally, it corresponds to the advancement of social justice. Companies may partake in philanthropy volunteer work or collaborations with profit organizations to address social challenges (Matten & Moon 2008).
3. *Environmental Responsibility;* Decreasing the impact through practices. Environmental responsibility involves reducing the effects of business operations, on the environment. There are various ways in which companies can promote sustainability like reducing carbon emissions handling waste responsibly and preserving resources. They can also use technologies, eco-friendly sourcing and sustainable products (Hart, 1995). The 3 R's (i.e. Re-duce/Re-use/Re-cyle) are considered an important principle and foundation of Environmental Responsibility.

It is important to be noted that among the three above mentioned Responsibility dimensions, there is some overlap and that is why it is extremely important that are studied as a whole and not separately (Scott & Gough, 2006). For this reason, the term Sustainable Development is academically considered an inclusive term as not only it describes the way that an organization responds to customers' expectations, but it also elaborates on the methodology used for the management of its available resources and its operation outcomes (Voukis, 2021).

8.2 Sustainability Reporting

In sustainability reporting practices, the concept of the Triple Bottom Line (TBL) goes beyond economic performance in order to include social and environmental aspects. According to a scholar publication (Elkington, 1997), it is suggested that a company's success should be measured not by profits but by its impact on society and the environment.

Measuring TBL involves monitoring performance indicators (KPIs) related to social and environmental factors. These indicators may cover areas such as revenue growth, carbon emissions reduction efforts, employee satisfaction levels and community involvement (Slaper & Hall 2011).

Social responsibility strategies may involve various initiatives such as;

- Ensuring ethical and environmentally responsible practices in the supply chain.
- Engaging in philanthropy by donating money or resources to causes.
- Implementing programs focused on sustainability to reduce carbon footprint and manage resources efficiently.

- Fostering employee engagement, through creating a workplace environment and encouraging volunteerism among staff members.
- Creating supply chains involves working with suppliers to ensure they adhere to ethical labor standards minimize their environmental impact and comply with regulations. This may include conducting audits, on suppliers offering training sessions and establishing goals for sustainability (Pagell & Wu 2009).
- Corporate philanthropy refers to giving and supporting community projects. This might involve donations of goods or services and programs where employees volunteer their time. Engaging in philanthropy helps strengthen ties with the community and improves a company's reputation (Seifert et al., 2003). The enhanced company brand, in turn, leads to a series of benefits which span from the low cost of capital and successful investors' attraction to low employee turnover, effective attraction of employees of better quality and enhanced customers' loyalty.

Environmental sustainability initiatives concentrate on reducing the footprint of business activities. This could involve projects aimed at improving energy efficiency reducing waste and investing in energy sources. Companies may also focus on creating products and packaging through innovation (Porter & van der Linde, 1995).

Employee engagement in social responsibility efforts promotes a workplace environment and boosts job satisfaction. Businesses can motivate employees to participate in volunteer activities offer opportunities, for growth and cultivate a culture that values sustainability and accountability (Glavas & Kelley 2014).

9 Relationship Between Digital Transformation and Corporate Sustainability and Responsibility

9.1 Improving CSR Through DT

Previous studies (Aghayari, J., Valmohammadi, Ch., & Alborzi, M., 2023; Omonijo & Zhang, 2025) showed that DT significantly affects an organization's sustainability aspects through operation, customer, business model, technology, workforce, and collaboration.

And how are the above communicated to the organization's stakeholders, general public and the regulators? One important channel is the **sustainability reporting** which presents an excellent opportunity for communication. Nowadays, sustainability reporting regulations pose a significant challenge for reporting professionals within the organizations, as they are struggling with the following growing challenges:

- the need to report as derived by the directives and laws that are increasingly adding more pressure
- the lack of a single, globally approved and preferred solution for reporting
- what data to report
- the multitude of reporting frameworks and languages to adopt
- the published communication/justification regarding any organization's failures, compared to last years' performance or/and compared to other competitors' performance
- the concerns over third-party data

- greenwashing.

It is important to be noted that **greenwashing**, which can be simply described as the release of incorrect or exaggerated information on behalf of the organization, related to its sustainability actions can be extremely damaging for the organization itself. In the organizations' effort to effectively avoid greenwashing and at the same time to respond to the above-mentioned reporting need, organizations are turning to a rapidly growing set of tools and technologies that allow them to improve their data collection process, their communication of the data, hence, the actual sustainability report itself.

The integration of transformation can greatly boost a company's CSR endeavors by offering the tools and capabilities to enhance transparency, efficiency and engagement. Key areas where DT aids CSR efforts include;

4. *Utilizing Data for Informed Decision Making:* Data analytics empower companies to assess and manage their social and environmental impact. Real time data insights can shed light on aspects like resource consumption, carbon footprints and supply chain practices (Omonijo & Zhang, 2025; Lee & Kim 2017).
5. *Enhancing Transparency and Reporting:* Digital platforms enable improved reporting standards and transparency levels. For instance, blockchain technology can establish records of transactions and processes ensuring credibility and reliability in CSR reporting (Saberi et al., 2019).
6. *Engaging with Stakeholders:* Utilizing social media platforms and other digital communication tools allows companies to interact effectively with stakeholders gather feedback and cultivate a sense of community and collaboration around CSR initiatives (Capriotti, 2011).
7. *Boosting Operational Efficiency:* Through automation technologies like IoT companies can optimize resource consumption minimize wastage and enhance efficiency in alignment with sustainability objectives (Jabbour et al., 2020).

9.2 Examples from the Retail Sector

Nowadays, various important organizations globally have utilized technology in order to enhance their CSR practices (Xi & Wang, 2024; Stellantis, 2024). Walmart (Walmart, 2021) can be defined as one successful example as it has effectively reorganized its supply chain. Walmart has achieved improved resource management and reduced waste by adopting data analytics and IoT technology. Walmart has managed not only to boost its internal supply chain but also to positively change the supply chain of its suppliers in order to further secure the sustainability of its own supply chain.

In the healthcare industry, various organizations adopt the use of cloud health records and telemedicine that not only enhance patient care but also reduce the environmental footprint linked to travel and paper based medical records.

Similarly, Siemens, in the manufacturing sector leverages IoT and AI to develop factories that prioritize energy efficiency and waste reduction showcasing a dedication to sustainability.

Within the financial services domain fintech advancements like banking and blockchain play a role in promoting financial inclusion and transparency while aligning with broader social responsibility objectives.

Challenges and opportunities emerge alongside this transformation journey. These include safeguarding data privacy and security in initiatives bridging the divide for more inclusive growth and managing the sustainability aspects of digital infrastructure, like data centers to address environmental concerns (Le Monde, 2024; Garcia & Lopez, 2023a, b).

9.3 More Important Topics that Require Action

1. *Data Privacy and Security:* It's essential to safeguard data in today's world. Businesses need to put in place cybersecurity measures and adhere to data protection laws to build trust and steer clear of issues (Cavoukian, 2012). As fraud and cyber hacking increases and constantly evolves with innovative tools, organizations struggle to implement robust data protection systems that can safeguard their data (Smith & Thompson, 2023a, b).
2. *Bridging the Digital Divide*: Ensuring access, to technologies is crucial for fostering inclusive growth (Nguyen & Lee, 2024). Past research explored effective strategies to bridge the digital divide by focusing on enhancing digital literacy and providing affordable technology access. While analyzing various initiatives, including community education programs and subsidized internet services, that have successfully increased digital inclusion, the study emphasized the importance of collaborative efforts between governments, private sectors, and non-profit organizations to ensure equitable access to digital resources. Companies can support literacy programs, offer affordable tech solutions and collaborate with governments and NGOs to bridge the gap (Chen & Zhang, 2024; Van Dijk, 2020; Patel & Singh, 2023a, b).
3. *Sustainable Digital Infrastructure:* Managing the impact of infrastructure like data centers is key. Businesses can embrace eco IT practices, utilize energy sources and deploy energy efficient technologies to minimize their carbon footprint (Harmon & Auseklis, 2009; Garcia & Lopez, 2023a, b). Previous research explored recent advancements in energy-efficient technologies that support sustainable digital infrastructure. While analyzing developments in low-power processors, green networking, and sustainable software engineering, it emphasized the role of these technologies in reducing energy consumption and promoting environmental sustainability in the digital sector (Kumar & Sharma, 2023a, b).

10 Conclusion

Digital Transformation and Corporate Sustainability Responsibility go hand in hand each complementing the other. By integrating cutting edge technologies companies can propel CSR efforts, for long term success and sustainability. In addition to operation, customers' satisfaction, business model, technology, workforce, and collaboration, the sustainability reporting process, which has become mandatory for many organizations and it is expected to become mandatory for more organizations in the near future, can be boosted through DT. Moreover, DT can contribute to the prevention of serious dangers/threats such as high cost of capital and greenwashing.

This study emphasizes the significance of taking an approach that integrates technology, engages stakeholders, promotes transparency and enhances efficiency in achieving sustainable business practices. It contributes to the discourse on DT and CSR by:

- Demonstrating how technology-driven strategies enhance corporate sustainability efforts.
- Identifying the role of digitalization in mitigating CSR risks such as greenwashing and financial instability.
- Emphasizing the growing importance of digital tools in mandatory sustainability reporting and compliance management.
- Highlighting how stakeholder engagement, transparency, and operational efficiency are strengthened through AI, blockchain, and big data analytics.

While this chapter aimed to shed some light to the intersection of DT and CSR, several critical areas such as the below permit and require further exploration:

- Ethical Challenges of Digital Sustainability
- Impact of Digitalization on Global ESG Regulations
- Role of Emerging Technologies in Green Innovation
- Measuring the ROI of Digital-Driven CSR
- Digital Inclusion and Sustainability

Future research could enhance the level of understanding of digital sustainability, helping the organizations to navigate evolving CSR landscapes with resilience while also securing innovation and differentiation. It should be highlighted that the organizations that effectively merge technological advancements with responsible corporate strategies will emerge as industry leaders, driving meaningful impact for both business and society.

References

Abraham, R., Schneider, J., Vom Brocke, J.: Data governance: a conceptual framework, structured review, and research agenda. Int. J. Inf. Manag. **49**, 424–438 (2019)

Aghayari, J., Valmohammadi, C., Alborzi, M.: The impact of digital transformation on sustainability: a case of the Iranian telecom industry. Int. J. Res. Ind. Eng. **12**(3), 306–320 (2023)

Aguirre, S., Rodriguez, A., Ortiz, F.: A practical approach to robotic process automation. In: Proceedings of the Future Technologies Conference, pp. 56–66 (2017)

Armbrust, M., et al.: A view of cloud computing. Commun. ACM **53**(4), 50–58 (2010)

Ashton, K.: That 'internet of things' thing. RFID J. **22**(7), 97–114 (2009)

Besson, P., Rowe, F.: Strategizing information systems-enabled organizational transformation: a transdisciplinary review and new directions. J. Strat. Inf. Syst. **21**(2), 103–124 (2012)

Bowers, C., Connolly, R.: The influence of digital transformation on cybersecurity: the case of small and medium-sized enterprises. J. Small Bus. Enterp. Dev. **27**(6), 1160–1181 (2020)

Brynjolfsson, E., McAfee, A.: The Second Machine Age: Work, Progress, and Prosperity in a Time of Brilliant Technologies. W.W. Norton & Company (2014)

Buhse, W.: Digital leadership: Navigating organizational transformation in the digital age. Int. J. Bus. Manag. **58**(4), 345–360 (2023)

Capriotti, P.: Communicating corporate social responsibility through the internet and social media. Corp. Commun. Int. J. **16**(3), 221–238 (2011)

Carroll, A.B.: Corporate social responsibility: evolution of a definitional construct. Bus. Soc. **38**(3), 268–295 (1999)

Carroll, A.B.: A history of corporate social responsibility: Concepts and practices. The Oxford Handbook of Corporate Social Responsibility (2008)

Cavoukian, A.: Privacy by design: Origins, meaning, and prospects. Privacy Protection Measures and Technologies in Business Organizations: Aspects and Standards, pp. 170–208 (2012)
Chen, Y., Zhang, X.: Data-driven strategies for competitive advantage: a comprehensive review. J. Strat. Inf. Syst. **29**(4), 301–318 (2024a)
Chen, Y., Zhang, X.: Public-private partnerships in addressing the digital divide: a global perspective. J. Technol. Soc. **29**(4), 301–318 (2024b)
Chen, Y., Roberts, K.: Enhancing cybersecurity compliance through employee training programs: a case study approach. J. Bus. Ethics **168**(4), 789–804 (2024)
Choudhury, A., Harding, J.A., Tiwari, M.K.: Data analytics for industrial engineering: exploring the impact of big data on manufacturing. Comput. Ind. Eng. **128**, 536–547 (2019)
Creusen, U.: Digital leadership: strategies for leading in the digital era. J. Digit. Bus. **12**(2), 99–115 (2024)
Davenport, T.H., Ronanki, R.: Artificial intelligence for the real world. Harv. Bus. Rev. **96**(1), 108–116 (2018)
Elkington, J.: Cannibals with Forks: The Triple Bottom Line of 21st Century Business. Capstone (1997)
Foss, N.J., Saebi, T.: Fifteen years of research on business model innovation: how far have we come, and where should we go? J. Manag. **43**(1), 200–227 (2017)
Friedman, M.: The social responsibility of business is to increase its profits. N. Y. Times Mag. **13**(1970), 32–33 (1970)
Gao, F., Zhang, L.: Power Thirst Complicates ESG Investors' Love Affair with Tech Stocks. Reuters (2024). https://www.reuters.com/sustainability/sustainable-finance-reporting/power-thirst-complicates-esg-investors-love-affair-with-tech-stocks-2024-09-26/
Garcia, M., Lopez, R.: Sustainable practices in data center management: reducing the carbon footprint. J. Green Comput. **18**(2), 89–105 (2023a)
Garcia, M., Lopez, R.: Predictive analytics in supply chain management: trends and applications. Int. J. Prod. Econ. **250**, 108–120 (2023b)
Garcia, M.E., Thompson, H.: Navigating GDPR and HIPAA compliance: challenges and strategies for healthcare providers. Health Inf. Manag. J. **52**(3), 200–215 (2023)
Glavas, A., Kelley, K.: The effects of perceived corporate social responsibility on employee attitudes. Bus. Ethics Q. **24**(2), 165–202 (2014)
Harmon, R.R., Auseklis, N.: Sustainable IT services: assessing the impact of green computing practices. In: Proceedings of the 2009 40th Annual Hawaii International Conference on System Sciences, pp. 1–10 (2009)
Hart, S.L.: A natural-resource-based view of the firm. Acad. Manag. Rev. **20**(4), 986–1014 (1995)
Ivanov, D., Dolgui, A., Sokolov, B., Ivanova, M., Potryasaev, S.: The impact of digital technology and Industry 4.0 on the ripple effect and supply chain risk analytics. Int. J. Prod. Res. **57**(3), 829–846 (2019)
Jabbour, C.J.C., Jabbour, A.B.L.S., Foropon, C., Filho, M.G.: When titans meet–Can industry 4.0 revolutionise the environmentally-sustainable manufacturing wave? The role of critical success factors. Technol. Forecast. Soc. Chang. **132**, 18–25 (2020)
Jiang, C.: Research on the impact of digital finance on high-quality development of enterprises. Am. J. Manag. Sci. Eng. **9**(1), 15–22 (2024). https://doi.org/10.54097/z0nyts98
Kamble, S.S., Gunasekaran, A., Dhone, N.C.: Sustainability through digital transformation: EU practices. Sustainability **15**(2), 456 (2023). https://doi.org/10.3390/su15020456
Kovynyov, I., Buerck, A., Mikut, R.: Design of transformation initiatives implementing organisational agility: an empirical study. J. Organ. Chang. Manag. **37**(2), 189–210 (2024)
Kumar, A., Sharma, P.: Customer behavior prediction using machine learning: a systematic review. J. Retail. Consum. Serv. **61**, 102–115 (2024a)
Kumar, A., Sharma, P.: Advancements in energy-efficient technologies for sustainable digital infrastructure. Int. J. Sustain. Technol. **15**(1), 67–82 (2024b)

Lee, S.M., Kim, S.T.: The impact of AI on CSR: challenges and opportunities. J. Bus. Res. **70**, 175–180 (2017)
Lee, J., Kim, H.: The moderating role of green operational innovation in the relationship between technological innovation, ESG, and corporate sustainability. Humanit. Soc. Sci. Communi. **11**, 389 (2024). https://doi.org/10.1057/s41599-025-04389-8
Lemon, K.N., Verhoef, P.C.: Understanding customer experience throughout the customer journey. J. Mark. **80**(6), 69–96 (2016)
Li, X., Wang, J., Zhou, Y.: An empirical assessment of the influence of digital transformation on the sustainable growth of sports enterprises. J. Bus. Res. **156**, 104258 (2023). https://doi.org/10.1016/j.jbusres.2023.104258
Li, F., Wang, J.: Risk management through predictive analytics: a case study in the financial sector. J. Risk Financ. **26**(1), 89–105 (2025)
Li, W., Chen, J.: Enterprise ESG performance, digital transformation, and firm innovation: evidence from Chinese listed companies. SAGE Open **14**(3), 215824402412916 (2024). https://doi.org/10.1177/21582440241291680
Liker, J.K.: The Toyota Way: 14 Management Principles from the World's Greatest Manufacturer. McGraw-Hill, New York (2004)
Lv, W.: Digital Transformation and Corporate Social Responsibility: Empirical Evidence from Chinese Listed Companies (2023)
Madnick, S.E., Johnson, S., Huang, K.T.: Managing cybersecurity resources: a cost-benefit analysis. Springer (2009)
Matten, D., Moon, J.: "Implicit" and "explicit" CSR: a conceptual framework for a comparative understanding of corporate social responsibility. Acad. Manag. Rev. **33**(2), 404–424 (2008)
Moskaliuk, J., Ott, A.: Leadership 4.0: adapting leadership styles in the era of digital transformation. J. Leadersh. Stud. **17**(1), 25–39 (2023)
Nakagawa, M.: Sustainability Is Falling on the CEO To-Do List. Customers Still See It as a Priority. The Wall Street Journal (2024). https://www.wsj.com/articles/sustainability-is-falling-on-the-ceo-to-do-list-customers-still-see-it-as-a-priority-bfb8d4d2
Nguyen, T.K., Lee, H.J.: The role of data governance in enhancing data security and usability. Int. J. Data Gov. **10**(1), 45–60 (2024)
Ni, X., Zhao, H., Wang, L.: Paradigm shift in sustainability disclosure analysis: empowering stakeholders with CHATREPORT, a language model-based tool. Artif. Intell. Bus. **12**(3), 224–238 (2023). https://doi.org/10.48550/arXiv.2306.15518
NIST. (2018). Cybersecurity Framework. National Institute of Standards and Technology. Retrieved from NIST Cybersecurity Framework
Nosratabadi, S., Atobishi, T., Hegedűs, S.: Social sustainability of digital transformation: empirical evidence from EU-27 countries. Adm. Sci. **13**(2), 78 (2023). https://doi.org/10.3390/admsci13020078
Nguyen, T.K., Lee, H.J.: Evolving cyber threats and the struggle for robust data protection systems. Int. J. Inf. Secur. **29**(1), 45–60 (2024)
O'Connor, D., Murphy, S.: The impact of predictive analytics on decision-making processes in healthcare. Health Inform. J. **29**(2), 159–172 (2023)
O'Connor, D., Williams, P.: Developing comprehensive security protocols to meet ISO 27001 standards in SMEs. J. Small Bus. Manag. **63**(1), 88–102 (2025)
Omonijo, O.N., Zhang, Y.: Examining the relationship between technological innovation, environmental social governance and corporate sustainability: the moderating role of green operational innovation. Humanit. Soc. Sci. Commun. **12**, 107 (2025). https://doi.org/10.1057/s41599-025-04389-8
Pagell, M., Wu, Z.: Building a more complete theory of sustainable supply chain management using case studies of 10 exemplars. J. Supply Chain Manag. **45**(2), 37–56 (2009)

Patel, R., Singh, S.: Leveraging data analytics for strategic decision-making in organizations. J. Bus. Anal. **22**(3), 201–215 (2023a)

Patel, R., Singh, S.: Strategies for bridging the digital divide: enhancing digital literacy and affordable technology access. J. Digit. Incl. **22**(3), 201–215 (2023b)

Patel, R., Lee, S.: The role of encryption technologies in maintaining data integrity across cloud platforms. Int. J. Inf. Secur. **29**(1), 45–60 (2024)

Pine, B.J., Gilmore, J.H.: The Experience Economy: Work is Theatre & Every Business a Stage. Harvard Business Press (1999)

Porter, M.E., Kramer, M.R.: Strategy and society: the link between competitive advantage and corporate social responsibility. Harv. Bus. Rev. **84**(12), 78–92 (2006)

Porter, M.E., van der Linde, C.: Toward a new conception of the environment competitiveness relationship. J. Econ. Perspect. **9**(4), 97–118 (1995)

Qing, H., Jin, L.: Does ESG and digital transformation affect corporate sustainability? The moderating role of green innovation. Sustainability **15**(9), 5432 (2023). https://doi.org/10.3390/su15095432

Raja, S.: Redefining success: the impact of transcendent leadership in business. J. Bus. Ethics **168**(4), 789–805 (2023)

Rigby, D.K., Sutherland, J., Noble, A.: Agile at scale. Harv. Bus. Rev. **94**(5), 88–96 (2016)

Ross, J.W., Sebastian, I.M., Beath, C.M.: How to develop a great digital strategy. MIT Sloan Manag. Rev. **58**(2), 7 (2016)

Russell, S., Norvig, P.: Artificial Intelligence: A Modern Approach. Pearson Education Limited, London (2016)

Saberi, S., Kouhizadeh, M., Sarkis, J., Shen, L.: Blockchain technology and its relationships to sustainable supply chain management. Int. J. Prod. Res. **57**(7), 2117–2135 (2019)

Schallmo, D.R.A., Williams, C.A., Boardman, L.: Digital transformation of business models—best practice, enablers, and roadmap. Digital Transformation Now! Springer (2017)

Schallmo, D.R., Williams, C.A., Boardman, T.: The role of digital technologies in corporate sustainability: a bibliometric review and future research agenda. J. Sustain. Sci. Manag. **19**(1), 112–129 (2024). https://doi.org/10.1007/s43621-024-00269-7

Sharma, I.: The 7 best practices of transcendent leadership. J. Leadersh. Organ. Stud. **30**(2), 123–137 (2023)

Schwaber, K., Sutherland, J.: The Scrum Guide. Scrum.org (2017)

Scott, W., Gough, S.: Sustainable development within UK higher education: Revealing (2006)

Seifert, B., Morris, S.A., Bartkus, B.R.: Comparing big givers and small givers: financial correlates of corporate philanthropy. J. Bus. Ethics **45**(3), 195–211 (2003)

Singh, A., Verma, R.: The impact of regular audits on maintaining cybersecurity standards in educational institutions. Comput. Secur. **120**, 102856 (2023)

Slaper, T.F., Hall, T.J.: The triple bottom line: what is it and how does it work? Indiana Bus. Rev. **86**(1), 4–8 (2011)

Smith, J.A., Thompson, L.M.: Data governance frameworks in modern enterprises: ensuring data quality and compliance. J. Inf. Manag. **35**(2), 123–137 (2023a)

Smith, J.A., Thompson, L.M.: Implementing robust cybersecurity measures to enhance data integrity and stakeholder trust. J. Cybersecur. Manag. **15**(2), 123–137 (2023b)

Smith, J.A., Nguyen, L.T.: Implementing advanced cybersecurity protocols to enhance data integrity in financial institutions. J. Cybersecur. Res. **15**(2), 101–115 (2023)

Stebbins, G.: Transcendent leadership: a modern approach to organizational excellence. Leadersh. Quart. **35**(1), 45–60 (2024)

Sund, K.J., Bogers, M., Villarroel, J.A., Foss, N.J.: Managing tensions between new and existing business models. MIT Sloan Manag. Rev. **61**(1), 8 (2020)

Tapscott, D., Tapscott, A.: Blockchain Revolution: How the Technology Behind Bitcoin is Changing Money, Business, and the World. Penguin (2016)

Tewes, S., Palmer, M., Hildebrandt, M.: Responsible digital transformation for a sustainable society. Technol. Soc. **75**, 102109 (2023). https://doi.org/10.1016/j.techsoc.2023.102109

Van Dijk, J.: The Digital Divide. Polity Press, Cambridge (2020)

Verhoef, P.C., et al.: Digital transformation: a multidisciplinary reflection and research agenda. J. Bus. Res. **122**, 889–901 (2021)

Verhoef, P.C., Kannan, P.K., Inman, J.J.: From multi-channel retailing to om-ni-channel retailing: introduction to the special issue on multi-channel retailing. J. Retail. **91**(2), 174–181 (2015)

Vial, G.: Understanding digital transformation: a review and a research agenda. Mana. Digit. Transform. 13–66 (2021)

Vogue Business (2024). Heritage made modern: Power to the Marketer lands in London. https://www.voguebusiness.com/story/events/heritage-made-modern-power-to-the-marketer-lands-in-london

Voukis, M.: Model for the integration of Sustainable Development in Higher Education Institutions: the importance of measuring the degree of its integration. Doctoral dissertation/thesis, Panteion University of Social and Political Sciences, Greece (2021)

Walmart. (2021). Walmart's 2021 ESG Report. Retrieved from Walmart ESG

Wang, L., Sun, Y.: How does corporate ESG performance affect sustainable development? The mediating role of green innovation. J. Clean. Prod. **395**, 136345 (2024). https://doi.org/10.1016/j.jclepro.2024.136345

Wessel, M., Levie, J., Siegel, E.A.: The problem with legacy ecosystems. Harv. Bus. Rev. **99**(1), 48–57 (2021)

Westerman, G., Bonnet, D., McAfee, A.: Leading digital: turning technology into business transformation. Harvard Business Review Press (2014)

Wickremasinghe, M., Gamage, A., Jayasundara, N.: Sustainable business decision modelling with blockchain and digital twins: a survey. Sustain. Digit. Innov. **8**(1), 57–76 (2024). https://doi.org/10.48550/arXiv.2405.12101

World Economic Forum (2024). Top 10 Emerging Technologies of 2024. https://www.weforum.org/publications/top-10-emerging-technologies-2024/in-full/methodology-emerging-technologies-2024/

Xi, L., Wang, H.: The influence of green innovation and digital transformation on the high-quality development of enterprises: the mediating role of ESG management. Sustainability **16**(24), 10923 (2024). https://doi.org/10.3390/su162410923

Yang, L., Li, J., Xu, D.: The impact of digital transformation on the sustainable development of manufacturing enterprises: mediating effect of green innovation. J. Clean. Prod. **315**, 132779 (2025). https://doi.org/10.1016/j.jclepro.2025.132779

Zhang, R., Zhou, Z.: A comparative study of information security risk management frameworks in cloud computing. J. Cloud Comput. **7**(1), 1–15 (2018)

Zhang, Y., Zhou, H., Chen, F.: The impact of digital transformation on corporate sustainability: new evidence from Chinese listed companies. Front. Environ. Sci. **10**, 1047418 (2022). https://doi.org/10.3389/fenvs.2022.1047418

Zhang, Y., Zhou, X.: Digital transformation and ESG performance: a quasinatural experiment based on the implementation of environmental protection law in China. J. Environ. Manag. **330**, 117123 (2024). https://doi.org/10.1155/2024/8895846

Zhang, Y., Li, W., Chen, X.: Digital transformation and corporate sustainability: the moderating effect of ambidextrous innovation. Systems **11**(7), 344 (2023). https://doi.org/10.3390/systems11070344

Paving the Way for Digital Transformation in Contemporary Organizations Through Building Strong Organizational Capabilities: Implications for Employee Engagement Toward Change

Sofia Chatzi(✉)

Department of Management Science and Technology, Athens University of Economics and Business, 76 Patission Str, 10434 Athens, GR, Greece
hatsof@aueb.gr

Abstract. Digital transformation is a prevailing trend in social and economic growth, offering substantial benefits to individuals and organizations globally. Today, there is a broad consensus that "*technology does matter*" because technology creates fundamentally new capabilities in organizations and people's lives and radically improves the organizational performance, operational efficiency, and sustainable development of organizations. However, despite the advantages of digital technologies, research indicates that many organizations encounter difficulties in realizing effective digital transformation because they tend to emphasize technological improvements without developing comprehensive plans and cohesive digital strategies that fully acknowledge the importance of the human element in these transformations. To help the leadership comprehend how cutting-edge technology and business models are interdependent and how this relationship affects the fundamentals of organizational behavior, a model of core foundational organizational capabilities is developed and presented in the following sections. Some preliminary research evidence indicates that, when handled effectively, these capabilities can foster employee engagement and adjustment to technological change. So, this study aims to highlight gaps in the existing literature and raise several important questions that scholars studying digital transformation and the role of employees in the change process need to explore with the goal of helping practitioners manage digital transformation more effectively within their organizations.

Keywords: Organizational Capabilities · Strategic-Human Resources Alignment · Digital Transformational Leadership · Cross-Functional Collaboration and Communication · Competent Human Resources · Digital Culture · Innovative and Agile Human Resource Management Practices

1 Introduction

In the twenty-first century, the business world is significantly impacted by the continuous introduction and advancement of revolutionary digital technologies. Some of the most disruptive digital technologies such as technology, artificial intelligence (AI), cloud,

A. K. Zarkada (Ed.): GMA-GAMMA 2025, SPBE, pp. 25–53, 2025.
https://doi.org/10.1007/978-3-031-96930-0_3

smartphone, social media, chatbots or virtual assistants, cybersecurity, and big data among others, undoubtedly have brought about numerous advantages to organizations, such as improved performance and efficiency, flexibility, diversity, automatization of processes and activities, reduced costs or brand image improvement or reputation. They have also offered various significant benefits, such as digital transformation (DT) and adaptability, leading to an improvement in employee and consumer experience and an increase in satisfaction.

Digitalization refers to the adoption of cutting-edge digital technology and solutions to modernize and enhance corporate processes (Li et al., 2016). The above statement describes one aspect of the concept of *digital transformation* where digitalization helps organizations utilize digital technologies to enhance their current operations by facilitating more efficient coordination and generating more consumer value through improved user experience (Pagani & Pardo, 2017). However, the rendering of the term *digital transformation* is incomplete unless we consider it a means of changing elements of the organization, such as structure, processes, and culture, to adapt to increasing market needs fueled by digital technologies (Nasiri et al., 2020).

Digital technologies have provided organizations with several advantages; however, research indicates that many organizations encounter difficulties in attaining successful digital transformation because they prefer to prioritize technology enhancements without developing comprehensive plans and cohesive digital strategies that fully recognize the importance of the human element in these transformations. Numerous empirical studies have demonstrated how firms undergo digital transformation and which organizational capabilities are critical to this process. Among them, information technology (IT), artificial intelligence, machine learning, and big data analytics constitute important technological, digital, and operational capabilities for promoting digital transformation (e.g., Grewal et al., 2001; Gordini & Veglio, 2017; Hausladen & Zipf, 2018; Nadkarni & Prügl, 2021), demonstrating a pronounced asymmetry of emphasis favoring technology over humans and teams in a digital work setting (Jedynak et al., 2021). On the contrary, little conceptual or empirical research has recommended key strengths that enable digital transformation aside from reforming internal processes in the areas of information technology and strategic management (see, for example, Hooi & Chan, 2023). From a managerial standpoint, organizations cannot simultaneously build strong technological, digital, and operational capabilities to ease the shift to the digital era unless they invest in developing essential foundational capabilities that aid employees in adapting to the upcoming digital transformation changes. Therefore, the lack of effective implementation of a digital workplace is strongly linked to inadequate change management procedures and the failure of top management to recognize that organizational change often stems from the insights and experiences of individuals who possess intimate knowledge of their daily operations (Tabrizi et al., 2019).

Literature has highlighted hindering factors in digital transformation, such as organizational culture, employees' perceptions and participation, and digital governance. Nonetheless, paramount among the impediments to failure or inadequate advancement in digital transformation is the unprepared organization (Gupta, 2018). Rigid structures, departmental silos, ineffective communication, excessive formalization, and strict rules can impede the overall efficiency of the organization (De Smet & Gagnon, 2018). Implementing change in an organization can be challenging due to employee resistance and lack of understanding regarding the necessity of change. Additionally, the success of change implementation depends on the employees' willingness to adopt new digital tools, which is influenced by their belief that these tools will require minimal effort and enhance job performance (e.g., Davis, 1989; Venkatesh & Davis, 2000). Furthermore, it might encounter challenges such as a lack of alignment between business needs and the skills of the current staff, leaders who are not effectively embracing change, and a lack of understanding about the necessity of human resource management strategies in promoting digital transformation effectiveness. Organizations that do not possess the necessary skills and competencies to effectively utilize these technologies are unable to generate new value or enhance existing value.

Digital transformation is a ubiquitous change process that impacts employees across all organizational levels, fundamentally alters job demands and regular duties, and compels people to reevaluate previous approaches to their work (e.g., Kensbock & Stockmann, 2020; Shoss, 2017). Research has shown that such changes modify job roles, resources, and expectations, and can emotionally affect employees, potentially damaging their engagement with the change process (Albrecht et al., 2020, p. 4). So, employees' role in the change process and how they are affected when planning for digital transformation should be largely considered (Buvat et al., 2018). Organizations often overlook this; Kaur & Sharma (2021) indicated that leaders continue to perceive the transformation process as primarily involving physical alterations rather than social transformations.

To achieve digital transformation, organizations need to focus on two essential elements: the integration of technologies across their value chain, and the implementation of changes that impact their workforce, culture, and expertise. Human resources (HR) serve as a competitive asset for any organization, and it is essential for organizations to align HR management with operations management to achieve optimal outcomes. The efficiency of digital transformation relies heavily on employees' willingness to not just adapt to but also actively support digital change (Oreg et al., 2011).

Several key inquiries require resolution. Based on having deeply analyzed the behaviors and cognitions of the individual actors during transformational change, how can leadership enable effective digital transformation in their organizations? What organizational capabilities are essential for facilitating employee engagement in the digital transformation process? Studying digital transformation through the lens of the organizational behavior (OB) perspective could extend our knowledge of how employees engage in the transformation process. Although workplace digitalization is vital for business survival, employee engagement is important for maintaining a competitive advantage (Hooi, 2021). Handling changes and stakeholder engagement effectively through applying strategic pillars that will support these aspects is essential for the success of a digital transformation project. Yet, limited research has explored how a unified set of change-related organizational capabilities affects engagement with change. Thus, the main aim of this study is to address the gap in the existing literature by identifying foundational organizational capabilities that activate employee enthusiasm for change and their active involvement in the achievement of successful digital transformation.

Furthermore, since digital transformation extends beyond technical dimensions and fundamentally involves human change that necessitates authentic support, a secondary question that needs to be addressed is how HR professionals can leverage the comprehensive understanding of the workforce and their cross-functional duties to synchronize their management with the technological initiatives and the overarching business objectives. To bridge the skills gap, for example, it is necessary to adopt new approaches that keep pace with the progress of digital technology. The HR specialists can help organizations surmount the obstacles of human resource development. By uncovering or suggesting novel approaches, technologies, or innovative strategies, they can empower leadership with the capacity to effectively address issues related to employee adjustment during digitalization. They can also inform and motivate the top management about new capabilities while helping stakeholders manage change to ensure seamless adoption throughout the organization. Thus, as digital transformation bestows a strategic role on core business functions, HR emerges as a strategic partner in the change management process by assessing employees' preparedness, promoting effective communication, delivering training, and analyzing the impacts of change. As a result, this study also tries to highlight the importance of HR in helping the top management redefine how it perceives the employees in the digital era and understand the meaning of crafting a digital organizational context.

In summary, behind the surface of technological improvements lies a vital foundation: the human factor. The ability to effectively manage organizational change is critical to the success of digital transformation, in addition to the adoption of new technology. Digital transformation as a faster change paradigm, creating a more adaptive, creative, and resilient organization, differs from traditional change management, which aims to move an organization from one stable state to another. Tackling change in digital transformation needs capabilities aimed at guiding enterprises through the complexity of transition while guaranteeing a seamless integration of technology, culture, and strategy. So, the proposed model advances the existing literature by developing a capabilities-based model for facilitating change engagement, utilizing the well-established job demands-resource (JD- R) model (Bakker & Demerouti, 2007) and dynamic capabilities theory (Teece, 2007; Teece et al., 1997), according to which employees' views and willingness to embrace workplace digitalization are primarily based on core organizational capabilities such as that proposed in this study, namely strategic-human resources alignment, a digital transformational leadership, cross-functional collaboration and communication, competent human resources, a digital culture, innovative and agile human resource management practices. It is considered one of the first to offer a valuable foundation for future research to further explore and expand the scope of workplace digitalization-change engagement areas, which are important but understudied themes. Finally, it gives practical advice to industry practitioners on what aspects of the workplace environment could keep employees engaged in the digital transformation process.

2 Literature Review and Model Development

2.1 Attitudes to Change: The Case of Employee Engagement Toward Change

As modern organizations continuously evolve and need to fast adjust to changes like pandemics, climate change, worldwide markets, competition, technological progress, etc. successful organizational change mostly hinges on the vision, planning, implementation, and communication strategies employed by top executives (Oreg et al., 2011); but it primarily depends on employees embracing new paradigms of thought and behavior (Choi, 2011). Nevertheless, in the context of digital transformation, changes typically concentrate solely on internal processes, neglecting or even disregarding the ramifications of such changes on individuals. Therefore, sustainable change management demands insights from appropriate theories and research regarding the most influential elements to promote positive employee attitudes toward change triggered by digital transformation.

Lines (2005) defines attitudes toward change as employees' thoughts, feelings, and behavioral intentions—whether positive, negative, or neutral—toward organizational change initiatives. Negative attitudes are often referred to as resistance to change (Jones & Van de Ven, 2016) or cynicism and pessimism about change (Andersson & Bateman, 1997). Positive attitudes are described through constructs like acceptance of change (Iverson, 1996), openness to change (Wanberg & Banas, 2000), change readiness (Armenakis et al., 1999), and affective commitment to change (Herscovitch & Meyer, 2002).

Several studies on organizational change have examined negative attitudes, such as employee resistance (e.g., Armenakis et al., 1999; 1993), by focusing on individual characteristics. These include personality traits (e.g., openness to change), cognitive processes (e.g., self-efficacy beliefs), and whether employees have access to external coping resources (e.g., timely information and participation in decision-making) to help them deal with change.

On the contrary, a large number of researchers have emphasized the importance of understanding positively oriented employee attitudes to change (e.g., Mathews & Linski, 2016). *Change engagement* as employee engagement (Bakker & Demerouti, 2014) is a constructive motivational factor that facilitates the active and positive participation of employees in transformation initiatives. Successful organizational change necessitates individuals who are invigorated by transformation, willing to experiment with new approaches, and who actively endorse and embrace new projects through modifications in their attitudes and behaviors (e.g., Armenakis et al., 1993). Nevertheless, engagement has not been thoroughly explored as a prerequisite of change adaptation, which is noteworthy given that it may provide the inspiration and motivation needed to handle change (van den Heuvel et al., 2017).

Employee engagement, on the other hand, has continued to be a focus of interest for the past 20 years, as employees who are passionate, motivated, and fully engaged are recognized as a key source of competitive advantage (Albrecht et al., 2015). Schaufeli and his colleagues (2002) describe engagement - a notion that continues to be prevalent in academic literature - as a positive, fulfilling state of mind that is related to work, characterized by vigor, dedication, and absorption (p. 74). The job demands-resources (JD-R) model (Bakker & Demerouti, 2007), on the other hand, is a prominent framework for explaining engagement. The model was primarily developed to integrate and elaborate upon two previously developed research traditions, the stressor-strain perspective and its theoretical basis for explaining the negative effects of job demands (e.g., job burnout) and the motivation perspective that has predominantly led the research on work engagement and organizational commitment (Bakker & Demerouti, 2014).

Job demands activate employees' physical and/or psychological (cognitive and emotional) energy and skills to respond to specific physical, psychological, social, or organizational aspects of the job that are related to certain physiological and/or psychological losses (Demerouti & Bakker, 2011). *Job resources*, on the other hand, pertain to the physical, psychological, social, or organizational elements of the work environment that can mitigate job demands and their detrimental health effects, facilitate the achievement of work objectives, and promote personal development and knowledge acquisition (Demerouti & Bakker, 2011). Empirical work has predominantly focused on researching job-level factors related to employee engagement, such as skill variety, task identity, task significance, autonomy, and feedback (e.g., Crawford et al., 2010). Apart from these factors, the job demands-resources (JD-R) model has indicated contextual factors as active in employees reaching their work goals. So, aspects of the organizations, including leadership support, clarity of organizational vision and objectives, culture and climate, organizational support, and supportive human resource (HR) management practices, can influence engagement both directly and indirectly (Albrecht et al., 2018).

Change engagement as an alternative aspect of employee engagement that applies to the context of change has been identified as a critical success factor for effective organizational change (see Albrecht et al., 2020). Albrecht and his colleagues (2020) have recently defined change engagement as a lasting and positive psychological state related to work, marked by genuine enthusiasm and a strong willingness to support, embrace, and advocate for organizational change (Albrecht et al., 2020; p. 4). Although engagement to change can be facilitated either by change-related organizational resources, job-related change resources, and change-related personal resources, it can also be prevented by change-related demands at all levels, including job, organization, etc. (Albrecht et al., 2020). For example, the need to provide employees with new skills or adjust the culture to facilitate digital transformation can adversely influence employee attitudes to change. Some preliminary research has highlighted the role of change-related personal resources such as change meaningfulness, change-related psychological safety, and change self-efficacy in change engagement (e.g., Albrecht et al., 2023). However, there is a lack of studies exploring how employees' perceptions of the organizational environment contribute to fostering engagement in the change process, although the need to understand the 'how' of adaptation to such changes in depth becomes more than a necessity. This contradicts the organizational change literature, which posits that analyzing organizational capabilities concerning change is crucial for comprehending the processes that facilitate successful change implementation (e.g., Detert et al., 2000).

The concept of *organizational capabilities* originates from the competitive advantage literature (Teece et al., 1997). This idea of competitive advantage is based on the resource-based view, which views an organization as a unique combination of various resources and competencies. More specifically, according to the resource-based view (RBV), firms can attain and sustain competitive advantage by developing resources, tangible and intangible assets such as HR skills (Galbreath, 2005), in such ways that are difficult to be imitated, and are rare (Barney, 1986). An organization's capability refers to its ability to carry out a series of coordinated tasks using its resources to achieve a specific outcome, while a resource is an asset or input to production, either tangible or intangible, that the organization owns, controls, or has semi-permanent access to (Helfat & Peteraf, 2003, p. 999). So, capabilities refer to an organization's capacity to utilize and combine resources, through organizational routines, to achieve its objectives.

Literature has highlighted two capabilities' classifications: the operational and the dynamic capabilities (Helfat & Peteraf, 2003). Operational capability is described as an advanced operation or set of routines that, along with their input flows, provide an organization's management with a range of decision-making opportunities to produce key deliverables (Winter, 2000, p. 983). It is only in recent years that scholars have started to focus on the capacities required to adapt to variations in the internal and external environment, or, more specifically, the capabilities required for change. Teece and his colleagues have called these capabilities dynamic capabilities (Teece et al., 1997). Dynamic capabilities refer to an organization's ability to refresh competencies to stay current with the changing business environment. In other words, they refer to the capacity of the organization to sense environmental changes, seize emerging opportunities, adapt, integrate, and reconfigure their existing resources is crucial for success in a disruptive (digital) world (Konopik et al., 2022; Helfat et al., 2009; Teece, 2007; Teece et al., 1997). These capabilities do not entail the creation of goods or the provision of services. On the contrary, they construct, integrate, or reorganize operational capabilities (Teece et al., 1997) and indirectly enhance the firm's output (Helfat & Peteraf, 2003).

Scholars dealing with the issue of understanding DT and what factors facilitate organizations in developing mechanisms for rapid adaptation to unexpected, repetitive, and ongoing changes have been referenced to specific organizational capabilities that are required to adapt to these fluctuations. Thus, this inspiring venue explores and theorizes organizational factors, including agile digital culture, alongside factors like digital business leadership and building digital talent (e.g., Carcary et al., 2016). Eby et al. (2000) presented preliminary empirical evidence supporting the potential relevance of larger contextual variables in the development of positive change attitudes. They discovered that employees who assessed their division's practices and procedures as flexible were more likely to rate their organization and its employees as more adaptable to change.

Organizational resources establish the framework within which employees perceive their work experiences (e.g., Leiter & Bakker, 2010). Thus, it is believed that the distally organizational capabilities may affect employees' engagement toward change through their perceptions of proximal work resources (see Albrecht et al., 2018). Taking from employee engagement literature and the relative research that has identified aspects of the organizations including leadership support, clarity of organizational vision and objectives, culture and climate, organizational support, and supportive human resource (HR) management practices that can affect engagement both directly and indirectly (Albrecht, et al., 2018; 2020) in the context of change, a model of organizational capabilities required during digital transformation with a special focus on the role of employees in the digital era, was developed and is presented below. A top-down approach to change management was chosen, as it is the most common method for managing organizational changes caused by digitalization. These transformations typically arise from the implementation of a new strategy at the organizational level. Nevertheless, the adoption of change-supportive behavior plays a crucial role in effectively carrying out changes (Rafferty et al., 2013). Consequently, examining the internal factors that influence positive employee attitudes towards change triggered by digital transformation, specifically, the factors that empower individuals to promote and engage in organizational change, is one of the priorities of leadership to help organizations become digital.

Despite the widespread development of digital applications, it is impossible to fulfill all of an organization's requirements using digital means; hence, human values cannot be ignored. When it comes to providing consumers with improved services and improving their level of empowerment, the mix of technology and human resources plays a significant role. The relevant literature in this area helped us generate and theorize some key foundational capabilities that act as a starting point for the digital transformation journey. These are: Strategic-Human Resources Alignment; Digital Transformational Leadership; Cross-Functional Collaboration and Communication; Competent Human Resources; Digital Culture; and Innovative and Agile human resource management practices.

A focused search on each factor was then performed, examining both theoretical frameworks and empirical research related to digital transformation and change management in organizational contexts. The exploration of digital transformation began broadly to gain insights into current research trends, followed by a more focused search to identify studies examining how organizational factors influence individual-level attitudes, specifically engagement in the digital transformation process.

2.2 Organizational Capabilities in the Digital Transformation Era

2.2.1 Strategic-Human Resources Alignment

What are the factors that contribute to the success or failure of digital transformation initiatives? Many organizations prioritize specific technology over the importance of integrating change into their overall business strategy, understanding the impact on employees, and establishing a system to keep leaders informed about adoption and employee engagement. On the contrary, best-practice organizations excel in the design and execution of digital platforms through the use of effective strategies, governance models, knowledge flow procedures, and technology.

An explicitly formulated digital transformation strategy is crucial for the overall success of implementing digital technologies. However, despite possessing a well-defined vision for a new corporate strategy, senior executives may encounter significant challenges or even insurmountable obstacles when it comes to implementing the necessary changes. To successfully implement a digital transformation program, it is necessary to effectively communicate the vision, articulate the reasons behind the change, and develop compelling documentation that outlines the impact of the transition and the advantages it brings to the employees. Strong vision and clear goals that are communicated to all employees enhance engagement, commitment, and investment in digital change. Research has indicated that effective vision communication which involves the top management shares the organization's strategic vision with employees, who then internalize it; the top management and employees work together to implement the vision, with the top management helping link employees' tasks to the vision; and employees, in turn, use the vision to guide their daily work and actions (Venus et al., 2019), inspires employees to promote the organization's sustained objectives and encourages them via daily tasks to attain the desired future state, which may eventually lead to engagement (Oreg & Berson, 2019).

Aligning, thus, employees with the digital strategy at large helps organizations improve communication between leaders and the rest of the business, maintain employee focus on strategic goals, especially if the personal goals align with the organization's purpose (Bass & Avolio, 1993), and build a culture of transparency. Furthermore, this alignment fosters a climate open to experimentation, innovation, and change, strengthening the organization's operational capability. Finally, it facilitates cross-functional teams to focus on mutual digital transformation goals and make compatible their different visions and plans.

Plain and consistent information regarding the reasons for the transition, the anticipated results, and how this transition will affect personnel, reduce uncertainty and build confidence between management and personnel. Engaging employees in open dialogues, where they may voice their opinions and seek clarification, encourages them to feel included and involved in the transformation process (Kotter, 2012). Allowing team members to engage in making and carrying out decisions promotes accountability as well as ownership. This strategy not only inspires employees but also harnesses their views and experiences to improve agile practices (Worley & Doolen, 2006).

Proposition 1. Strategic-human resources alignment is positively related to employee engagement toward change.

2.2.2 Digital Transformational Leadership

Digital transformation encompasses far more than the mere adoption of novel technologies. A comprehensive strategy that spans the entire organization, from top to bottom, is necessary. It necessitates alterations in culture, procedures, and mentality. It requires a readiness to question the existing state of affairs, engage in experimentation, and derive knowledge from failures. Furthermore, it necessitates effective leadership.

Effective leadership has a crucial role in achieving success in digital transformation (Canterino et al., 2020; Weber et al., 2022). Nevertheless, the tumultuous alterations that come with an organization's digital transformation make leadership a demanding and intricate undertaking (Cortellazzo et al., 2019). Research indicates that leaders require behavioral complexity to effectively navigate a dynamic environment and successfully tackle hard tasks associated with change (Weber et al., 2022). On one side, leaders should demonstrate task-oriented behaviors to accomplish organizational goals while also giving guidance and instructions to their subordinates. These behaviors encompass actions taken by executives who are digitally oriented and demonstrate initiative in reimagining the structure of the organization and formulating a digital strategy (Tabrizi et al., 2019).

Leaders are responsible for establishing the vision, clearly communicating the plan, and motivating the organization to adopt change enthusiastically. They should possess a willingness to take risks and derive knowledge from their failures. Furthermore, they must furnish the required resources, assistance, and instruction to empower the organization to achieve success. Leaders with digital transformational capabilities often show transformative vision and governance and create cultures open to digital initiatives to maximize business benefits.

On the other hand, they need to possess the skills to recognize, comprehend, and consider not just their employees' personal needs and worries but also their strengths and shortcomings to offer suitable assistance for digital transformation. Leaders who exhibit people-oriented behaviors provide help and demonstrate empathy towards their employees, build trustworthy working relationships with them, show gratitude and trust in their capabilities, acknowledge their efforts, and assist in their skill development (Yukl, 2012). People-oriented leadership behaviors in digital transformation (DT) encompass a range of actions aimed at assisting employees in managing the challenging tasks in their work, convincing them of the significance and importance of these transformative organizational changes, and fostering the growth of their skills to enhance their talent (Kane et al., 2019).

When leaders demonstrate inspiration and provide personalized assistance (Piccolo & Colquitt, 2006), they can effectively engage employees who may be resistant to participating in the change process. They enhance cooperation among them to facilitate the achievement of shared objectives. They also foster a sense of hope and dedication in employees, motivating them to work towards accomplishing the desired transformation. Employees who are supporters and have a willingness to embrace change demonstrate a strong belief in the urgent need to actively engage in the process of implementing change (e.g., Holt et al., 2010).

Therefore, leaders can facilitate the success of organizations in the digital age by (1) keeping up with emerging technologies and exploring potential opportunities in digital development; (2) establishing the course of digital transformation and investment strategy; and (3) guiding the team to adapt quickly and accurately (Swift & Lange, 2018). Undoubtedly, the latter characteristic has been proven mandatory as transformation needs leaders who help and support employees during the process and provide coaching that helps them succeed in their work. Facilitating communication and fostering open relationships that promote dialogue are considered essential, alongside acknowledging and treating employees as experts in their respective fields. So, leaders must possess a high level of emotional and social intelligence to effectively handle organizational change (e.g., Cortellazzo et al., 2019; Usai et al., 2020). Additionally, they should have strong networking and communication skills (e.g., Sousa & Rocha, 2019; Verhoef et al., 2021). Promoting active engagement and involvement in the process of change can reduce the likelihood of opposition to IT or, conversely, enhance the probability of IT acceptance. Employees who have personally invested in the design of the workplace are likely to support transformational initiatives more easily (e.g., Meske & Junglas, 2021).

Proposition 2. Digital transformational leadership is positively related to employee engagement toward change.

2.2.3 Cross-Functional Collaboration and Communication

Organizations undergoing digital transformations require the cooperation and collaboration of individuals from different functional areas. This is because these synergies provide greater advantages than individuals working alone, including enhanced idea generation, greater flexibility in developing solutions, and a means of crossing boundaries to foster innovation within the organization. Setting explicit objectives offers team members a clear framework and ensures that cross-functional teams remain focused on shared responsibilities and desired results (e.g., Santa et al., 2010). This, in turn, fosters a greater degree of cross-functional collaboration (Pinto et al., 1993). Effective collaboration across different functions is seen as an essential component of agility and ambidexterity, both of which are critical for advancing digital transformation (DT). Agile structures are considered to have a greater ability to recognize business opportunities and dangers at an early stage and implement appropriate adjustments due to their adaptable character, enabling rapid revisions (Chan et al., 2019; Leso et al., 2023; Teece et al., 2016).

Communication and collaboration can take place through direct in-person interactions as well as through digital tools (e.g., Mesmer-Magnus & DeChurch, 2009). The increased virtuality through a range of cutting-edge and innovative collaboration technologies helps improve both task-related and interpersonal communication among members from different functions, leading to positive outcomes in collaborative work. These include effective knowledge sharing, smooth information flow, rapid and accurate task coordination, and greater transparency in work processes (e.g., Anders, 2016).

However, cross-functional cooperation can be unsettled. Different stakeholders in an organization hold distinct assumptions, expectations, knowledge, and perceptions regarding technical breakthroughs and digital transformations, as well as the methods to attain them (Santa et al., 2010). The presence of diverse team members may lead to miscommunications or misunderstandings, thus necessitating careful analysis and consideration while setting transformation goals and objectives, as these factors can impact the outcomes. Existing in-person collaboration and direct communication practices often lead to resistance or difficulty adapting to the shift toward digitalized work and communication processes (e.g., Hur et al., 2019). Establishing bridges, cooperation, and co-creation is crucial for facilitating the success of individuals with different cultures or backgrounds in adopting a shared language and defining and implementing precise and consistent goals.

Thus, to facilitate digital transformation, the organization must transition from its conventional, hierarchical framework to a flexible, decentralized structure where cross- functional teams bring together members from different functional areas to work collaboratively. Collaboration provides significant benefits to all parties. It provides access to new resources and promotes mutual connection and communication, enabling engagement and, finally, the realization of common goals (Schmutz et al., 2019).

Proposition 3. Cross-functional collaboration and communication is positively related to employee engagement toward change.

2.2.4 Competent Human Resources

Highly competent human resources are considered crucial capital in the digital transformation era. According to Mattajang (2023), skilled human resources not only possess the required technical skills but also can maintain a state of continuous learning and innovation. For individuals to not only survive in an unpredictable working environment but also to make significant contributions in producing value for businesses and society as a whole, creativity and innovation are essential components that must be present.

Therefore, organizations compelled to use these cutting-edge technologies must prioritize the enhancement of employees' skills and knowledge to optimize their performance (e.g., Chatterjee et al., 2022). It is claimed that the use of new technologies significantly impacts the performance of employees and is considered the primary factor influencing their willingness to adopt new technologies, hence facilitating digital transformation (e.g., Xu et al., 2013). On the other hand, it is argued that digital technologies by themselves have little or no effect on an organization's performance. On the contrary, digital literacy, namely the skills, knowledge, and abilities of employees used while interacting with digital technologies (see Kozanoglu & Abedin, 2021), is also important, as it constitutes a critical dynamic capability of organizations during their digital transformations that strengthens operational capabilities. Individuals who possess a strong desire to learn, excellent problem-solving skills, adaptability to change, and the necessary expertise to address digital transformation challenges (Butschan et al., 2019) are a prerequisite for tackling the challenges raised during transformation (Blanka et al., 2022). Organizations need to equip their workforce with digital skills and other competencies – intrapreneurial competency which is characterized by encouraging opportunities; proactivity; flexibility; drive; and willingness to take risks (Vargas-Halabí et al., 2017) is among them – to meet the organizational goals and reap the benefits of their expenditures in technological advancements (Blount et al., 2016; Kane et al., 2019a).

Digital transformation in the realm of workforce and skill development entails a significant shift in culture and mindset, challenging the existing norms, fostering innovation, and embracing continuous learning even in the face of setbacks (Varshney, 2020). This cultural shift facilitates the provision of appropriate skill development opportunities to the workforce at the appropriate moment. Regularly acquiring new skills to stay relevant is crucial for the success of an organization throughout times of change. Insufficient digital literacy among employees might pose significant challenges in adapting to digital transformations. If employees are not prepared to use digital technology applications, such as chatbots, as active and engaged participants in their daily activities, employee experience practices will fail to create a personalized, compelling, and memorable environment for employees (Kozanoglu & Abedin, 2021). According to Bawden and Robinson (2009), this situation is more likely to create an excessive amount of information and feelings of anxiousness.

Research indicates that individuals who perceive themselves as competent in technology usage are likely to engage with technology not only in conventional manners but also through innovative exploration. Thus, research emphasis has shifted from the study of 'use' to investigating the 'effective use' (Burton-Jones & Grange, 2008). Enhancing employee confidence in organizations, either through group training sessions during implementation or personalized coaching afterward, could be an effective method of encouraging workers to not only embrace digital workplace technologies and engage in the transformation process but also experiment with technology in innovative ways. Furthermore, the enhanced use of the technology enhances the employees' sense of its utility (e.g., Gefen et al., 2003), hence promoting the employees' deliberate digital preparedness (see Höyng & Lau, 2023). If employees have the belief that their skills can be enhanced and developed through training, and if they can gain proficiency in technology and utilize it, they will willingly embrace the task of digitally transforming both the organization and themselves (e.g., Jones et al., 2021).

Proposition 4. Competent human resources is positively related to employee engagement toward change.

2.2.5 Digital Culture

Organizational culture refers to the broader pattern of an organization's mores, values, and beliefs. It is less conscious and not so directly visible to employees (Schneider et al., 1994). The organization's culture can be inferred from what the employees say and believe about why organizations do what they do and focus on what they focus on (e.g., "We value serving our customers"). Culture is grounded in history and tradition and is a source of collective identity and commitment (e.g., Schein, 1985).

Every prosperous organization possesses a fundamental culture. The core culture is crucial to the organization's functioning, serving as the central hub for its operational success. The alignment of the organization's strategy and fundamental leadership practices with its core culture is crucial. Any management concept or innovative approach, regardless of its quality, will not be effective if it is not aligned with the prevailing culture. If the culture is not in line with and supportive of the plan of the organization, even the most excellent strategy will either come to a halt or be unsuccessful. Thus, alignment is crucial for the performance of any organization. In the absence of it, focus is lost, and individuals, structures, and procedures operate in conflicting directions.

The theoretical literature on organizational culture has extensively examined the impact that the founder and upper management have on shaping an organization's culture. The leaders of an organization establish the fundamental culture, either intentionally or unintentionally, based on their background, inherent characteristics, socialization experiences, and understanding of the requirements for success in the market.

Organizations, on the other hand, are said to possess distinct cultures when their members collectively hold a shared set of assumptions, values, and beliefs. According to Schein (1985), culture is believed to be passed down through myths and stories. When a significant number of people within an organization understand and share the significance of these myths, a culture is considered to be present. Culture is primarily concerned with understanding the underlying causes and significance of events or phenomena.

Within the context of digital transformation, possessing a congruent culture enables organizations to execute this transition with efficiency and effectiveness and assure engagement toward change (e.g., Myilswamy & Gayatri, 2014; Rashmi, 2015). An inclusive culture establishes a norm for employees' involvement in the decision-making process, allowing them to freely voice their thoughts, exercise their judgment, and make spontaneous decisions (Shahzad et al., 2017). Furthermore, a culture that enables digital transformation is not solely focused on technologically proficient organizations.

Instead, it revolves around a digital mentality and digital habits developed within an agile, dynamic, collaborative, and creative work environment. Digital transformation necessitates a culture that not only supports the change but also facilitates the organization's overall strategy, a culture that encourages innovation, creativity, and an entrepreneurial mindset among employees by nurturing a tolerance for mistakes, management support for generating new ideas and lifelong learning, risk-taking and experimentation, competitiveness, and finally constructive conflict management as central organizational values (e.g., Bärenfänger & Otto 2015; Karimi & Walter, 2015). Transparency and effective communication are essential to get everyone motivated and engaged in the potential of new technology. Openness and transparency alleviate worries and ambiguity, providing a lucid perspective and facilitating comprehension. Participation and involvement provide the opportunity to actively participate, change, and take responsibility (Foerster-Metz et al., 2018). Flat hierarchies, on the other hand, and decentralized decision-making are also permitted in digital cultures (Shin et al., 2023). So, when employees perceive a digitally innovative culture as advantageous, a reciprocal organization-employee relationship is formed, leading employees to respond to the positive work environment, often through increased engagement (Saks, 2006).

Proposition 5. Digital culture is positively related to employee engagement toward change.

2.2.6 Innovative and Agile Human Resource Management (HRM) Practices

Teece and his colleagues (2016, p. 17) define agility as an organization's ability to quickly and effectively reallocate its resources to activities that create, protect, and capture value in response to changing internal and external conditions.

Amidst the current business landscape characterized by digital disruption, organizations that possess agility significantly surpass their competitors. Agile organizations are outcome-based, effectively handle challenges such as evolving skill demands, recruitment, and retention in a competitive talent market, and provide flexibility and interconnectedness among the internal stakeholders (Varshney, 2020). In other words, agile organizations possess a responsive HR function that swiftly and efficiently addresses changes and problems related to business requirements, workplace disruptions, and employee expectations and attitudes. Agile HR enhances the overall agility of the organization by advancing it via the alignment of personnel and processes.

Therefore, an imminent challenge for organizations in the upcoming years is to establish a set of agile and innovative practices for their staff that will streamline the process of digital transformation. Implementing Human Resource Management (HRM) practices in the digital age can serve as a mechanism for achieving the desired objectives of the digital strategy by exerting influence over and altering the skills, behaviors, and attitudes of employees.

Employee involvement with the digital workplace transformation can be gauged by cognitive, emotional, and behavioral reactions, including thoughts, feelings, and behaviors (Kaur & Sharma, 2021). As such, employees play a major role in determining whether a change is successful or not in the end (Neves et al., 2017). The majority of transformation initiatives fall short because people do not recognize the need for change.

Employee reactions might range from helpful and upbeat to antagonistic and downright hostile (e.g., Vakola & Nikolaou, 2005). Therefore, organizations should understand the current situation and use specific HR procedures to modify employees' approaches to change (Vakola, 2016). Employees' positive reactions to change, such as employee engagement to change, are influenced by several factors, including the availability of resources, open communication, innovative training and development programs, and strong supervisory support, among others.

Literature has identified various HRM practices favorable to DT, such as teamwork or promoting employee autonomy and team cohesion (Nicolas-Agustin et al., 2021). Organizations attain sustainable competitive advantage when human resource management practices align with organizational strategy amid digital transformation. HRM practices such as hiring people with the required skills, teleworking, promoting the involvement of employees to achieve the goal of digital transformation, augmenting digital competencies and skills, prioritizing cultural transformation and leadership development, and managing talent are very important. Furthermore, HRM practices that promote relationships and joint efforts, that support self-initiative and grant employees the freedom to start their own projects, can create an atmosphere that enhances creativity and personal accountability. However, to achieve agility, increasing persistent learning becomes a priority for organizations. This entails developing a learning atmosphere that encourages employees to actively pursue knowledge expansion.

More specifically, raising the intensity of persistent learning is a proactive strategy that can assist organizations in adapting to the rapid changes that the digital world brings. Organizations that fail to develop internal competencies may experience a decline in their competitive advantage in the future (Hess et al., 2016). Providing employees with active support for training and continuing education, as well as actively creating a learning environment that is supportive of learning, is one effective method for reaching this goal. Employees can continuously develop new competencies, such as creativity, proactivity, and risk-taking, when they are provided with opportunities to learn in an atmosphere that encourages the exchange of knowledge and ideas and when various training programs, including online platforms, workshops, seminars, etc., are made available to them. Research evidence suggests that both personal resources, like resilience, and organizational resources, such as access to information and training, enhance technology acceptance. The findings emphasize the importance of offering all employees opportunities for knowledge and training to support digital transformation while preserving their motivation and engagement in the process (Molino et al., 2020).

Studies have also shown that employees are more receptive to organizational change and engage in it when they comprehend the rationale behind it and perceive it as advantageous and favorable (Nemanich & Keller, 2007). Thus, research has demonstrated that leaders who effectively share and clarify the digital vision, specifically through digital transformation-oriented leadership behaviors, show the capacity to build trust, provide direction, and tackle any potential challenges that may arise during the digital transformation process, establish a foundation for enhancing employees' comprehension of the necessity for organizational change (Nemanich & Keller, 2007). Engaging in such behaviors can help employees create knowledge and accept the change strategy and its aims, ultimately decreasing their resistance to change (Zhao et al., 2016). The meta- analytical study conducted by Peng et al. (2021) showed that leaders can diminish employees' resistance to change by presenting an optimistic vision of the future. So, HR practices that engage leaders in continuous learning and personal growth, ensuring their capability to adjust to technological advances and effectively guide their employees towards the success of the digital vision, should be among the HR professionals' main priorities to enhance employee engagement in the change process.

One of the main objectives of HRM is the management of the organizational value system (culture). Change does not happen by implementing new mission statements, giving speeches, distributing newsletters, hosting a grand event to introduce a new approach, or even by altering the organization's structure (Schneider et al., 1996). Change can only occur and be sustained by modifying policies, practices, procedures, and routines (HRM practices) and demonstrating agility (HR agility), which in turn affects the beliefs, values, and attitudes that influence employee actions. Digital transformation necessitates a shift in the fundamental culture of the organization, and one effective approach to do this is by altering the artifacts, stated beliefs/values, and underlying assumptions (i.e., modifying the content). Individuals possess a strong inclination towards progress and advancement, and they demonstrate creativity, a crucial attribute in the process of digital transformation, when provided with such possibilities. Individuals like interpersonal communication, whether it is with colleagues or higher-ranking individuals, which makes the formal and informal aspects of these connections a prominent characteristic of organizational existence. To operate efficiently and achieve organizational performance, individuals require trust, support, and cooperation (Schneider et al., 1996). On the other hand, zero- sum, win-lose rivalries within organizations are not beneficial and can potentially harm operational efficiency and overall organizational performance. When HR executives learn to operate and introduce practices under assumptions like the above, they strengthen the operational capability of the organization, can function more effectively, and impact the beliefs, values, and eventually employee engagement that guide employee actions in a digital world.

Thus, fostering a culture of teamwork that promotes cooperation and knowledge-sharing is essential for establishing an adaptive and responsive organizational environment. Contemporary firms are progressively acknowledging the significance of efficient collaboration as essential for navigating business dynamics successfully. In the established teamwork culture, collaboration is perceived not merely as a collective work process but also as a concerted endeavor to attain a shared objective. A culture that promotes knowledge-sharing fosters continuity in organizational learning. Team members not only fulfill their responsibilities but also exchange experiences, knowledge, and insights with one another. This approach enables teams to learn from one another, comprehend diverse perspectives, and develop more imaginative solutions to difficulties.

In addition, even though the advent of novel technologies and solutions has significantly altered business models, production processes, as well as methods of working and collaborating, and technology has facilitated increased mobility, accelerated decision-making and delivery, and bolstered greater standardization, customization, and prosperity, nevertheless, the rapid advancement of digitalization has resulted in an intricate digital work setting within numerous organizations, potentially placing high demands on staff. The process of digitalization often results in an increased amount of information in the workplace. As the number of information and communication technologies grows, employees are expected to handle a greater volume of information. This poses a problem for the well-being of employees due to the overwhelming volume of information and information and communication technologies (ICTs) employed in the workplace, which can be difficult to handle effectively. The growing utilization of ICTs and the corresponding demand for information processing have a significant impact on work pace, leading to fragmentation, interruptions, information overload, and the continuous requirement for multitasking. Consequently, these factors result in an increased workload, elevated stress levels, heightened frustration, impaired memory, reduced ability to concentrate, intensified time constraints, and diminished control over one's tasks (for a review, see Trennery et al., 2021). When job demands are excessive and not supported by sufficient resources, it can result in poorer health, an increased risk of burnout, and lower levels of work engagement and well-being. On the other hand, higher subjective well- being can foster optimism and self-efficacy, which boosts task persistence and facilitates learning. This, in turn, leads to improved performance over time, greater resilience, and better adaptability in the face of digital transformation (Tenney et al., 2015). As a result, HR professionals must be conscious of the effects, whether advantageous or disadvantageous, and address them from the outset by implementing initiatives that improve employees' subjective well-being through providing them with employment resources. These resources are components of the job that assist individuals in managing demanding job requirements, achieving work objectives and performance, and fostering professional development. Job resources, such as feedback, work variety, assistance, and autonomy, have been illustrated by Bakker and Demerouti (2017). Automating benefits and leaving the administration to chatbots and virtual assistants, on the other hand, can reduce the levels of stress. Chatbots and virtual assistants in HRM improve not only the convenience of access but also the range of information that can be retrieved (Roberts et al., 2020), making the workplace more engaging and encouraging. Meske and Junglas (2021) found that employees' expectations of autonomy, competence, and connection in the digital workplace were associated with greater support for digital transformation.

The implementation of digital technology in the workplace allows employees to telecommute, providing them with the opportunity to remain in the comfort of their own homes while fulfilling their job responsibilities for their employers (Chatterjee et al., 2023). Occasionally, it grants them ample time to spend with their family and for personal pursuits. The inherent pleasure derived from this experience aids employees in effectively utilizing digital technology. Organizations embrace digital transformation to promote flexible working, prioritize productivity over hours worked, routinely assess workloads, and more. These techniques can assist personnel in organizations in properly managing their work-life balance while maintaining high levels of engagement, productivity, and efficiency. Nevertheless, scholars have concluded that teleworking influences engagement in both a positive and negative way. Specifically, teleworking could enhance engagement through increased job autonomy and reduced role conflict; however, if we want to minimize its negative effects on engagement, it must be reinforced through increased support and feedback and decreased role ambiguity (Sardeshmukh et al., 2012).

On the other hand, academics and HR practitioners frequently consider digital HRM as a widely applicable management solution that can expedite digital transformation.

The implementation of digital HRM in organizations can accelerate various HR processes, including e-recruitment, online training and development - utilizing webinars and virtual learning make re-skilling and up-skilling processes more efficient - (e.g., Zehir et al., 2020), online selection process (facilitating interview sessions for candidates, particularly during the Covid-19 pandemic), as well as reward and recognition and performance appraisal. Implementing a digital HR platform in an organization improves the efficiency of the HR processes and enables the organization to respond quickly and flexibly to changes. Effective onboarding and socialization by providing employees with instant responses and support are essential in helping new personnel become incorporated into the organizational culture and speeding up their contribution and engagement to organizational objectives. Furthermore, harnessing big data analytics in HR can help organizations gain greater insight into employee behaviors and attitudes during transformation, allowing HR to implement initiatives that improve employee engagement and retention (George et al., 2014). Thus, HR analytics have become a ubiquitous tool for HR managers, serving as a data-driven approach to enhance decision-making in areas such as recruiting, turnover, retention of both current and potential personnel, succession planning, and talent acquisition.

Indeed, as employees are widely regarded as the most precious resource in any organization due to their potential to significantly impact the critical performance metrics and the overall competitiveness of the organization, the ability of the organization to recruit and retain talented, skilled, and technologically savvy employees who will facilitate digital transformation through engagement to the process constitutes the most crucial factor for its survival and success in the digital era. Thus, nowadays, the recruitment practices of organizations have moved beyond the traditional methods of newspaper ads, employee referrals, career offices, and employment agencies to attract qualified applicants to more radical ones such as job boards, organization websites, and applicant tracking systems (Ployhart et al., 2017). Jackson and Schmidt (2023) draw attention to the appearance of talent intelligence platforms that utilize big data and predictive analytics to increase the efficiency of the recruitment process. Using insights generated by data, organizations can identify applicant profiles that align not only with current employment requirements but also with future business objectives (e.g., digital transformation), delivering a sustainable and more engaged talent stream. These new methods have become a necessity because individuals have altered the way they search for a job. Notably, young people search for employment information through the web or various sources such as blogs and electronic bulletin boards (Nikolaou et al., 2019). Innovative e-recruitment practices, such as digital screening methods, which enable organizations to assess the information the job seekers put online (e.g., social media), could be effective tools for organizations in the disruptive digital world.

In conclusion, in their study, Bouaziz and Hachicha (2018) found that implementing strategic human resource management practices, such as staffing, training, and compensation, can enhance employees' knowledge and foster their understanding of the overall operating environment. This, in turn, enables them to develop adaptive strategies to effectively respond and engage in disruptive situations. So, human capital is recognized as the primary catalyst for achieving agile outcomes, and human resource management practices play a crucial role in enhancing overall organizational performance. By implementing inventive frameworks and advanced HRM strategies, organizations may establish a solid basis for adapting and thriving in the digital age. It is crucial, however, to recognize that the HR management strategy throughout digital transformation is a continuous process that does not end after a single occurrence. Digital transformation is an ongoing process that requires organizations to regularly assess and adjust their strategies in response to technological breakthroughs and changes in business operations. This involves monitoring corporate trends, keeping up with state-of-the-art technologies, and regularly updating human resources strategies. Within the framework of an organization's digital transformation, implementing a thorough and long-lasting strategy for preparing and managing the workforce in the digital age is crucial for effective human resource management. In other words, by utilizing technology, data analytics, and customized techniques, organizations can attract top talent, promote cultural alignment, and speed up the incorporation of recruits, thereby driving engagement and resilience in an era of constant change.

Proposition 6. ***Innovative and agile human resource management practices is positively related to employee engagement toward change.***

3 Discussion

Digital transformation refers to a significant and disruptive change in an organization's culture, achieved by implementing new IT infrastructure and introducing new digital skill sets. This transition can be overwhelming for both personnel and the organization. Empirical evidence on how organizations are digitally transformed and what organizational capabilities are decisive for engaging employees in the digital transformation process is largely missing. Scholars have proposed many antecedents of employee engagement, such as job characteristics, perceptions of organizational support, appreciation and recognition, procedural fairness, and distributive justice (Saks, 2006). However, important questions remain unanswered: What are the key organizational drivers of employee engagement that support digital transformation, and how do their perceptions of the organizational elements impact their engagement in this process? By giving consideration to the role of these aspects in employee attitudes and, finally, behavior, the top management, with the assistance of HR professionals, can develop strategies for engagement and retention in digital transformation.

In a digital transformation context, it is suggested that strategic-human resources alignment, digital transformational leadership, cross-functional collaboration and communication, competent human resources, a digital culture, and innovative and agile human resource management practices constitute crucial capabilities for fostering employees' willingness to engage in the transformation process of their organizations and for boosting their ability to adapt and respond to changes the digital transformation brings to their work environment. So, this study contributes to the literature by considering organizational capabilities that go beyond technological aspects and suggesting elements such as people that are affected by the digital transformation and only peripherally considered by the relative literature (Henriette et al., 2015). It also indirectly highlights the importance of HRM as a strategic partner in enabling organizational change in the area of new technologies by apportioning resources, encouraging professional learning and development, and providing competent leadership and employees (e.g., Li & Herd, 2017; Chuang & Graham, 2018).

4 Practical Implications

The proposed model could help practitioners create a shared, easily communicated language for managing change due to technology penetration into employees' daily work activities. Additionally, it could serve as a valuable diagnostic tool for organizations seeking to understand and effectively manage strategies to harmonize employee attitudes toward change. Using the model of organizational capabilities as a basis for conducting in-house surveys, organizations could collect data about how employees are experiencing elements of the organization in the process of digital transformation and what interventions can facilitate engagement into change. More generally, the model reinforces the importance of organizations to develop change capabilities through change-conducive leadership, cultures, agile HRM practices, and effective team-based communication, among others.

More specifically, the effectiveness of the digital strategy is not exclusively determined by the choice of technology used. On the contrary, it hinges on the organization's capacity to invest in pertinent digital skills that are in line with both the digital strategy and the workforce. Aligning employees with the overarching digital strategy enhances communication between leadership and broader organization, sustains emphasis on strategic objectives, and fosters a culture of transparency and adaptability. As digital transformation involves significant changes in the operational processes, leaders must effectively handle these shifts as a vital necessity and ultimately consider and tackle through support the social and psychological consequences of digital change on their personnel to assure engagement to change.

Furthermore, organizations with a digital culture open to change and transparency in decision-making can help employees grow their capabilities and develop creative solutions to cope with digital disruptions (Kane et al., 2017). A culture of continuous improvement, continuous delivery, and communication enhances operational effectiveness by increasing employees' engagement and willingness to work together in teams, being motivated, flexible, gifted, self-disciplined, and organized, and showing a remarkable ability to improvise (Stettina & Heijstek, 2011). Culture, as a pivotal facilitator of innovation, establishes the necessary circumstances for attracting individuals who are well-suited to the organization and guarantees that individuals possessing the appropriate talents can flourish within the cultural framework. This, in turn, fosters a strong dedication to embracing change.

To maintain relevance and competitiveness in the global market, on the other hand, organizations must be vigilant and comprehend the dynamics of the business environment, monitor current trends, and implement measures to adopt HRM practices that enhance employee engagement in the digital era. HR as a strategic partner can implement optimal human resource management practices, such as providing training, hiring new employees with expertise in integration processes, or outsourcing specialized skills and competencies that are difficult to find (Hess et al., 2016), among others that are in line with the digital strategic objectives of the organization and have a positive effect on employees' responses to change, ultimately enhancing their level of engagement (Nitschke et al., 2020).

To conclude, we might have tried to change our organizations by changing their structures and processes to facilitate digital transformation, while the people are those who need changing. With changes in how people see digital transformation, the necessary changes in structure and process will occur.

5 Future Research

This study provides a review of the relevant literature on digital transformation and what organizational factors might affect employees' positive attitudes toward digital transformation. However, it cannot cover all the existing and emerging literature. As it also includes the author's interpretation of the relative literature and research findings without being confirmed through primary research, future research is encouraged to validate the identified foundational capabilities through both qualitative and quantitative methods. Additionally, the employee engagement to change construct has just recently been introduced as a context-specific concept of employee engagement applied during change (see Albrecht et al., 2020). So, it necessitates validation studies to focus on indicating its discriminant validity from existing measures of openness to change, willingness to change, and resistance to change.

Bringing employees in line with the organization's digital vision and values, while fostering a positive work environment through effective leadership, investing in learning and development for upskilling and reskilling, promoting teleworking (e.g., Bennett & McWhorter, 2021), focusing on employee engagement and motivation, allowing, and involving employees in the strategic decision-making of the organization, or creating a culture of teamwork, experimentation, and risk-taking conform to practices that the HR must adopt in the digital age. Therefore, redesigning HR practices that directly or indirectly convey the organization's digital vision and mission to the employees and facilitate a successful digital workplace transformation while maintaining a significant employee engagement level should be the focus of future research (see also Albrecht et al., 2020).

Additionally, another potentially fruitful avenue of future research is to identify the necessary leadership skills to accurately perceive and efficiently manage change, reduce opposition, and promote a collective commitment and engagement to the required changes. Internal contextual elements, such as the trustworthiness of the management, and leadership attributes like building trust, trying to develop teams, and leading by example, play a crucial role in determining the preparedness for change (e.g., Oreg et al., 2011).

Regardless of its quality, any HRM strategy will not be effective in practice if it is not aligned with the organizational culture. If the culture of an organization is not linked with and supportive of its HRM strategy for encouraging digital transformation, the strategy is likely to either stagnate or fail, as noted by Schneider (2000). So, future research should extend our knowledge of how organizational culture and HR practices interact in building organizational capacity in the digital transformation journey by boosting employee engagement.

In conclusion, a more detailed review and validation of the key capabilities for digital transformation, along with examining their interconnections, would be helpful in developing a comprehensive framework or roadmap for what factors affect employees engage in digital transformation process. Current organizational behavior (OB) models analyzing and forecasting human attitudes and behaviors during workplace digital transformation aim to provide valuable insights into the factors influencing individuals at various levels of analysis. While research at the individual level seems to dominate the field of change management, and some multilevel research has been conducted at the organizational level, studies on how work groups can affect individuals adjusting to digital transformation would supplement our understanding of this field.

References

Albrecht, S.L., Bakker, A.B., Gruman, J.A., Macey, W.H., Saks, A.M.: Employee engagement, human resource management practices and competitive advantage. J. Organ. Eff. People Perform. **2**, 7–35 (2015)

Albrecht, S., Breidahl, E., Marty, A.: Organizational resources, organizational engagement climate, and employee engagement. Career Dev. Int. **23**, 67–85 (2018)

Albrecht, S.L., Connaughton, S., Foster, K., Furlong, S., Yeow, C.J.L.: Change engagement, change resources, and change demands: a model for positive employee orientations to organizational change. Front. Psychol. **11**, 531944 (2020). https://doi.org/10.3389/fpsyg.2020.531944

Albrecht, S.L., Furlong, S., Leiter, M.P.: The psychological conditions for employee engagement in organizational change: test of a change engagement model. Front. Psychol. **14**(January), 1071924 (2023). https://doi.org/10.3389/fpsyg.2023.1071924

Anders, A.: Team communication platforms and emergent social collaboration practices. Int. J. Bus. Commun. **53**(2), 224–261 (2016). https://doi.org/10.1177/2329488415627273

Andersson, L.M., Bateman, T.S.: Cynicism in the workplace: some causes and effects. J. Organ. Behav. **18**, 449–469 (1997). https://doi.org/10.1002/(SICI)1099-1379(199709)18:5<449::AID-JOB808>3.0.CO;2-O

Armenakis, A.A., Harris, S.G., Field, H.S.: Making change permanent: a model for institutionalizing change interventions. In: Pasmore, W.A., Woodman, R.W. (eds.), Research in organizational change and development, pp. 97–128. JAI Press, New York, NY (1999). https://doi.org/10.1016/s0897-3016(99)12005-6

Armenakis, A.A., Harris, S.G., Mossholder, K.W.: Creating readiness for organizational change. Human Relations **46**, 681–703 (1993). https://doi.org/10.1177/001872679304600601

Bakker, A.B., Demerouti, E.: The job demands-resources model: state of the art. J. Manag. Psychol. **22**, 309–328 (2007)

Bakker, A.B., Demerouti, E.: Job demands-resources theory. In: Chen, P.Y., Cooper , C.L. (eds.), Wellbeing: A Complete Reference Guide, (Vol. III, pp. 37–64). Wiley- Blackwell, Chichester (2014)

Bakker, A.B., Demerouti, E.: Job demands–resources theory: taking stock and looking forward. J. Occup. Health Psychol. **22**, 273–285 (2017)

Bärenfänger, R., Otto, B.: Proposing a capability perspective on digital business models. In: Proceedings of the 2015 IEEE 17th Conference on Business Informatics, CBI '15, Washington, DC, USA, pp. 17–25 (2015)

Barney, J.B.: Strategic factor markets: expectations, luck, and business strategy. Manag. Sci. **32**(10), 1231–1241 (1986)

Bass, B., Avolio, B.J.: Transformational leadership and organizational culture Public Adm. Q. **17**, 112–121 (1993)

Bawden, D., Robinson, L.: The dark side of information: overload anxiety and other paradoxes and pathologies. J. Inf. Sci. **35**(2), 180–191 (2009)

Bennett, E.E., McWhorter, R.R.: Virtual HRD's role in crisis and the post Covid-19 professional lifeworld: accelerating skills for digital transformation. Adv. Dev. Hum. Resour. **23**(1), 5–25 (2021)

Blanka, C., Krumay, B., Rueckel, D.: The interplay of digital transformation and employee competency: a design science approach. Technol. Forecast. Soc. Chang. **178**, 121575 (2022). https://doi.org/10.1016/j.techfore.2022.121575

Blount, Y., Abedin, B., Vatanasakdakul, S., Erfani, S.: Integrating enterprise resource planning (SAP) in the accounting curriculum: a systematic literature review and case study. Acc. Educ. **25**(2), 185–202 (2016)

Bouaziz, F., Smaoui Hachicha, Z.: Strategic human resource management practices and organizational resilience. J. Manag. Dev. **37**(7), 537–551 (2018). https://doi.org/10.1108/jmd-11-2017-0358

Burton-Jones, A., Grange, C.: Using information systems effectively: a representational perspective. In: Proceedings of JAIS Theory Development Workshop. Sprouts: Working Papers on Information Systems, vol. 8, no. 21, pp. 1–35 (2008)

Butschan, J., Heidenreich, S., Weber, B., Kraemer, T.: Tackling hurdles to digital transformation – The role of competencies for successful industrial Internet of Things (IIoT) implementation. Int. J. Innov. Manag. **23**(4), Article 1950036 (2019). https://doi.org/10.1142/S1363919619500361

Buvat, J., et al.: The Digital Culture Challenge: Closing the Employee-Leadership Gap. Capgemini Digital Transformation Institute (2018)

Canterino, F., Cirella, S., Piccoli, B., Shani, A.B.: Leadership and change mobilization: the mediating role of distributed leadership. J. Bus. Res. **108**, 42–51 (2020). https://doi.org/10.1016/j.jbusres.2019.09.052

Carcary, M., Doherty, E., Conway, G.: A dynamic capability approach to digital transformation – a focus on key foundational themes. Nr Reading: Acad Conferences Ltd. (2016)

Chan, C.M.L., Teoh, S.Y., Yeow, A., Pan, G.: Agility in responding to disruptive digital innovation: case study of an SME. Inf. Syst. J. **29**(2), 436–455 (2019). https://doi.org/10.1111/isj.12215

Chatterjee, S., Chaudhuri, R., Vrontis, D., Giovando, G.: Digital workplace and organization performance: moderating role of digital leadership capability. J. Innov. Knowl. **8**(1), 100334 (2023). https://doi.org/10.1016/j.jik.2023.100334

Chatterjee, S., Chaudhuri, R., Vrontis, D., Basile, G.: Digital transformation and entrepreneurship process in SMEs of India: a moderating role of adoption of AI-CRM capability and strategic planning. J. Strateg. Manag. **15**(3), 416–433 (2022). https://doi.org/10.1108/JSMA-02-2021-0049

Choi, M.: Employees' attitudes toward organizational change: a literature review. Hum. Resour. Manag. **50**, 479–500 (2011)

Chuang, S., Graham, C.M.: Embracing the sobering reality of technological influences on jobs, employment and human resource development: a systematic literature review. Eur. J. Train. Dev. **42**, 400–416 (2018). https://doi.org/10.1108/ejtd-03-2018-0030

Crawford, E.R., Lepine, J.A., Rich, B.L.: Linking job demands and resources to employee engagement and burnout: a theoretical extension and meta-analytic test. J. Appl. Psychol. **95**(5), 834–848 (2010)

Cortellazzo, L., Bruni, E., Zampieri, R.: The role of leadership in a digitalized world: a review. Front. Psychol. **10**, Article e1938 (2019). https://doi.org/10.3389/fpsyg.2019.01938

Davis, F.D.: Perceived usefulness, perceived ease of use, and user acceptance of information technology. MIS Q. **13**(3), 319–340 (1989). https://doi.org/10.2307/249008

Demerouti, E., Bakker, A.B.: The job demands-resources model: challenges for future research. South Afr. J. Ind. Psychol. **37**, 1–9 (2011)

De Smet, A., Gagnon, C.: Organizing for the age of urgency. McKinsey (2018)
Detert, J.R., Schroeder, R.G., Mauriel, J.J.: A framework for linking culture and improvement initiatives in organizations. Acad. Manag. Rev. **25**(4), 850–863 (2000)
Eby, L.T., Adams, D.M., Russell, J.E.A., Gaby, S.H.: Perceptions of organizational readiness for change: factors related to employees' reactions to the implementation of team-based selling. Human Relations **53**(3), 419–442 (2000). https://doi.org/10.1177/0018726700533006
Foerster-Metz Ulrike S., Marquardt K., Golowko N., Kompalla A., Hell C.: Digital transformation and its implications on organizational behavior. J. EU Res. Bus. (2018). https://doi.org/10.5171/2018.340873
Galbreath, J.: Which resources matter the most to firm success? An exploratory study of resource-based theory. Technovation **25**(9), 979–987 (2005)
Gefen, D., Karahanna, E., Straub, D.W.: Trust and TAM in online shopping: an integrated model. MIS Q. **27**(1), 51–90 (2003). https://doi.org/10.2307/30036519
George, G., Haas, M.R., Pentland, A.: Big data and management. Acad. Manag. J. **57**(2), 321–326 (2014)
Goran, J., Laberge, L., Srinivasan, R.: Culture for Digital Age. McKinsey Quarterly (2017)
Gordini, N., Veglio, V.: Customers churn prediction and marketing retention strategies. An application of support vector machines based on the AUC parameter- selection technique in B2B e-commerce industry. Ind. Mark. Manag. **62**, 100–107 (2017). https://doi.org/10.1016/j.indmarman.2016.08.003
Grewal, R., Comer, J.M., Mehta, R.: An investigation into the antecedents of organizational participation in business-to-business electronic markets. J. Mark. **65**(3), 17–33 (2001). https://doi.org/10.1509/jmkg.65.3.17.18331
Gupta, S.: Organizational barriers to digital transformation. Doctoral Dissertation, KTH (2018)
Hakanen, J.J., Bakker, A.B., Schaufeli, W.B.: Burnout and work engagement among teachers. J. Sch. Psychol. **43**(6), 495–513 (2006). https://doi.org/10.1016/j.jsp.2005.11.001
Hausladen, I., Zipf, T.: Competitive differentiation versus commoditisation: the role of big data in the European payments industry. J. Paym. Strategy Syst. **12**(3), 266–282 (2018)
Helfat C.E., Finkelstein S., Mitchell W., Peteraf, M., Singh H., Teece D., Winter S.G.: Dynamic Capabilities: Understanding Strategic Change in Organizations. John Wiley & Sons, Hoboken (2009)
Helfat, C.E., Peteraf, M.A.: The dynamic resource-based view: capability lifecycles. Strateg. Manag. J. **24**(10), 997–1010 (2003)
Henriette, E., Feki, M., Boughzala, I.: The shape of digital transformation: a systematic literature review. In: Mediterranean Conference on Information Systems (MCIS) (2015)
Herscovitch, L., Meyer, J.P.: Commitment to organizational change: extension of a three-component model. J. Appl. Psychol. **87**, 474–487 (2002). https://doi.org/10.1037/0021-9010.87.3.474
Hess, T., Benlian, A., Matt, C., Wiesbock, F.: Options for formulating a digital transformation strategy. MIS Q. Exec. **15**, 123–139 (2016)
Holt, D.T., Helfrich, C.D., Hall, C.G., Weiner, B.J.: Are you ready? How health professionals can comprehensively conceptualize readiness for change. J. Gen. Intern. Med. **25**(1), 50–55 (2010)
Hooi, L.W.: Leveraging human assets for MNCs performance: the role of management development, human resource system and employee engagement. Int. J. Hum. Resour. Manag. **32**(13), 2729–2758 (2021)
Hooi, L.W., Chan, A.J.: Does workplace digitalization matter in linking transformational leadership and innovative culture to employee engagement? J. Organ. Chang. Manag. **36**(2), 197–216 (2023). https://doi.org/10.1108/JOCM-06-2022-0184
Höyng, M., Lau, A.: Being ready for digital transformation: how to enhance employees' intentional digital readiness. Comput. Hum. Behav. Rep. **11**, 100314 (2023). https://doi.org/10.1016/j.chbr.2023.100314

Hur, J.-Y., Cho, W., Lee, G., Bickerton, S.H.: The "smart work" myth: how bureaucratic inertia and workplace culture stymied digital transformation in the relocation of South Korea's capital. Asian Stud. Rev. **43**, 691–709 (2019). https://doi.org/10.1080/10357823.2019.1663786

Iverson, R.: Employee acceptance of organizational change: the role of organizational commitment. Int. J. Hum. Resour. Manag. **7**, 122–149 (1996). https://doi.org/10.1080/09585199600000121

Jackson, S.E., Schmidt, A.M.: Agile talent management: navigating organizational transformation. Acad. Manag. J. **66**(1), 5–22 (2023). https://doi.org/10.5465/amj.2017.1228

Jedynak, M., Czakon, W., Kuźniarska, A., Mania, K.: Digital transformation of organizations: what do we know and where to go next? J. Organ. Chang. Manag. **34**(3), 629–652 (2021)

Jones, M.D., Hutcheson, S., Camba, J.D.: Past, present, and future barriers to digital transformation in manufacturing: a review. J. Manuf. Syst. **60**, 936–948 (2021). https://doi.org/10.1016/j.jmsy.2021.03.006

Jones, S.L., Van de Ven, A.H.: The changing nature of change resistance: an examination of the moderating impact of time. J. Appl. Behav. Sci. **52**(4), 482–506 (2016). https://doi.org/10.1177/0021886316671409

Kane, G., Palmer, D., Phillips, A., Kiron, D., Buckley, N.: Achieving digital maturity, MIT Sloan Manag. Rev. 1–31 (2017)

Kane, G.C., Phillips, A.N., Copulsky, J., Andrus, G.: How digital leadership is(n't) different. MIT Sloan Manag. Rev. **60**(3), 34–39 (2019)

Kane, G.C., Phillips, A.N., Copulsky, J.R., Andrus, G.R.: The Technology Fallacy: How People Are the Real Key to Digital Transformation. MIT Press, Boston, MA (2019)

Karimi, J., Walter, Z.: The role of dynamic capabilities in responding to digital disruption: a factor-based study of the newspaper industry. J. Manag. Inf. Syst. **32**, 39–81 (2015). https://doi.org/10.1080/07421222.2015.1029380

Kaur, S., Sharma, R.: Emotion AI: integrating emotional intelligence with artificial intelligence in the digital workplace. In: Innovations in Information and Communication Technologies (IICT-2020), pp. 337–343. Springer, Cham (2021)

Kensbock, J.M., Stockmann, C.: "Big brother is watching you": surveillance via technology undermines employees' learning and voice behavior during digital transformation. J. Bus. Econ. **91**(4), 565–594 (2020). https://doi.org/10.1007/s11573-020-01012-x

Konopik, J., Jahn, C., Schuster, T., Hoßbach, N., Pflaum, A.: Mastering the digital transformation through organizational capabilities: a conceptual framework. Digit. Bus. **2**(2), 100019 (2022). https://doi.org/10.1016/j.digbus.2021.1000

Kotter, J.P.: Leading Change. Harvard Business Review Press, Boston (2012)

Kozanoglu, D.C., Abedin, B.: Understanding the role of employees in digital transformation: conceptualization of digital literacy of employees as a multi-dimensional organizational affordance. J. Enterp. Inf. Manag. **34**(6), 1649–1672 (2021)

Leiter, M.P., Bakker, A.B.: Work engagement: introduction. In: Bakker, Leiter, A.B., M.P. (eds.). Work Engagement: A Handbook of Essential Theory and Research. Psychology Press, New York, NY (2010)

Leso, B.H., Cortimiglia, M.N., Ghezzi, A.: The contribution of organizational culture, structure, and leadership factors in the digital transformation of SMEs: a mixed- methods approach. Cogn. Technol. Work **25**(1), 151–179 (2023)

Lewin, K.: Field Theory in Social Science, p. 6. Harper and Row, New York, NY (1951)

Li, F., Nucciarelli, A., Roden, S., Graham, G.: How smart cities transform operations models: a new research agenda for operations management in the digital economy. Prod. Plan. Control **27**(6), 514–528 (2016)

Li, J., Herd, A.M.: Shifting practices in digital workplace learning: an integrated approach to learning, knowledge management, and knowledge sharing. Hum. Resour. Dev. Int. **20**, 185–193 (2017). https://doi.org/10.1080/13678868.2017.1308460

Lines, R.: The structure and function of attitudes toward organizational change. Hum. Resour. Dev. Rev. **4**, 8–32 (2005). https://doi.org/10.1177/1534484304273818

Matarazzo, M., Penco, L., Profumo, G., Quaglia, R.: Digital transformation and customer value creation in Made in Italy SMEs: a dynamic capabilities perspective. J. Bus. Res. **123**, 642–656 (2021)

Mathews, B., Linski, C.M.: Shifting the paradigm: reevaluating resistance to organizational change. J. Organ. Chang. Manag. **29**, 963–972 (2016). https://doi.org/10.1108/JOCM-03-2016-0058

Mattajang, R.: Organizational agility: HR practices for facing rapid change in the digital era. Jurnal Info Sains: Informatika dan Sains **13**(3), 829–837 (2023)

Meske, C., Junglas, I.: Investigating the elicitation of employees' support towards digital workplace transformation. Behav. Inf. Technol. **40**(11), 1120–1136 (2021). https://doi.org/10.1080/0144929x.2020.1742382

Mesmer-Magnus, J.R., DeChurch, L.A.: Information sharing and team performance: a meta-analysis. J. Appl. Psychol. **94**, 535–546 (2009). https://doi.org/10.1037/a0013773

Molino, M., Cortese, C.G., Ghislieri, C.: The promotion of technology acceptance and work engagement in industry 4.0: from personal resources to information and training. Int. J. Environ. Res. Public Health **17**, 2438 (2020). https://doi.org/10.3390/ijerph17072438

Myilswamy, N.P., Gayatri, R.: A study on employee engagement: role of employee engagement in organizational effectiveness. Int. J. Innov. Sci. Eng. Technol. **1**(6), 331–333 (2014)

Nadkarni, S., Prügl, R.: Digital transformation: a review, synthesis and opportunities for future research. Manag. Rev. Q. **71**(2), 233–341 (2021). https://doi.org/10.1007/s11301-020-00185-7

Nasiri, M., Ukko, J., Saunila, M., Rantala, T.: Managing the digital supply chain: the role of smart technologies. Technovation 96–97 (2020)

Nemanich, L.A., Keller, R.T.: Transformational leadership in an acquisition: a field study of employees. Leadersh. Q. **18**(1), 49–68 (2007). https://doi.org/10.1016/j.leaqua.2006.11.003

Neves, P., Almeida, P., Velez, M.J.: Reducing intentions to resist future change: combined effects of commitment-based HR practices and ethical leadership. Hum. Resour. Manag. **57**(1), 249–261 (2017). https://doi.org/10.1002/hrm.21830

Nicolas-Agustin, A., Jimenez-Jimenez, D., Maeso-Fernandez, F.: The role of human resource management practices in the implementation of digital transformation. Int. J. Manpow. **43**(2), 1–26 (2021)

Nikolaou, I., Georgiou, K., Bauer, T.N., Truxillo, D.M.: Applicant reactions in employee recruitment and selection. The role of technology. In: Landers, R.N. (ed.) The Cambridge Handbook of Technology and Employee Behavior, pp. 100–130. Cambridge University Press, Cambridge, United Kingdom (2019)

Nitschke, C., Vallo Hult, H., Bigolin, F.: Shared workspaces of the digital workplace: from design for coordination to coordination for flexible design. In: Proceedings of the 53rd Hawaii International Conference on System Sciences (2020, January)

Oreg, S., Berson, Y.: Leaders' impact on organizational change: bridging theoretical and methodological chasms. Acad. Manag. Ann. **13**(1), 272–307 (2019)

Oreg, S., Vakola, M., Armenakis, A.: Change recipients' reactions to organizational change. J. Appl. Behav. Sci. **47**, 461–524 (2011)

Pagani, M., Pardo, C.: The impact of digital technology on relationships in a business network. Ind. Mark. Manag. **67**, 185–192 (2017)

Peng, J., Li, M., Wang, Z., Lin, Y.: Transformational leadership and employees' reactions to organizational change: evidence from a meta-analysis. J. Appl. Behav. Sci. **57**(3), 369–397 (2021). https://doi.org/10.1177/0021886320920366

Piccolo, R.F., Colquitt, J.A.: Transformational leadership and job behaviors: the mediating role of core job characteristics. Acad. Manag. J. **49**, 327–340 (2006). https://doi.org/10.5465/AMJ.2006.20786079

Pinto, M., Pinto, J., Prescott, J.: Antecedents and consequences of project team cross-functional cooperation. Manag. Sci. **39**(10), 1281–1297 (1993)

Ployhart, R.E., Schmitt, N., Tippins, N.T.: Solving the supreme problem: 100 years of selection and recruitment. J. Appl. Psychol. **102**(3), 291–304 (2017)

Rafferty, A.E., Jimmieson, N.L., Armenakis, A.: Change readiness. A multilevel review. J. Manag. **39**(1), 110–135 (2013)

Rashmi, R.: Building employee engagement through organizational culture: an empirical study of Indian IT industry. Prestig. Int. J. Manag. Res. **8**(2/1), 15–20 (2015)

Rich, B.L., Lepine, J.A., Crawford, E.R.: Job engagement: Antecedents and effects on job performance. Acad. Manag. J. **53**(3), 617–635 (2010)

Roberts, C., Kundavaram, R.R., Onteddu, A.R., Kothapalli, S., Tuli, F.A., Miah, M.S.: Chatbots and virtual assistants in HRM: exploring their role in employee engagement and support. NEXG AI Rev. Am. **1**(1), 16–31 (2020)

Saks, A.M.: Antecedents and consequences of employee engagement. J. Manag. Psychol. **7**(4), 600–619 (2006)

Santa, R., Ferrer, M., Bretherton, P., Hyland, P.: Contribution of cross-functional teams to the improvement in operational performance. Team Perform. Manag. **16**(3/4), 148–168 (2010). https://doi.org/10.1108/13527591011053241

Sardeshmukh, S.R., Sharma, D., Golden, T.D.: Impact of telework on exhaustion and job engagement: a job demands and job resources model. N. Technol. Work. Employ. **27**(3), 193–207 (2012). https://doi.org/10.1111/j.1468-005X.2012.00284.x

Schaufeli, W.B., Bakker, A.B.: Job demands, job resources and their relationship with burnout and engagement: a multi-sample study. J. Organ. Behav. **25**, 293–315 (2004)

Schaufeli, W.B., Salanova, M., Gonzalez-Roma, V., Bakker, A.B.: The measurement of engagement and burnout: a two-sample confirmatory factor analytic approach. J. Happiness Stud. **3**(1), 71–92 (2002)

Schein, E.H.: Organizational Culture and Leadership. Jossey-Bass, San Francisco (1985)

Schmutz, J.B., Meier, L.L., Manser, T.: How effective is teamwork really? The relationship between teamwork and performance in healthcare teams: a systematic review and meta-analysis. BMJ Open **9**(e028280), 1–16 (2019). https://doi.org/10.1136/bmjopen-2018-028280

Schneider, W.E.: Why good management ideas fail: the neglected power of organizational culture. Strategy Leadersh. **28**(1), 24–29 (2000)

Schneider, B., Brief, A.P., Guzzo, R.A.: Creating a climate and culture for sustainable organizational change. Organ. Dyn. **24**(4), 7–19 (1996). https://doi.org/10.1016/s0090-2616(96)90010-8

Schneider, B., Gunnarson, S.K., Niles-Jolly, K.: Creating the climate and culture of success. Organ. Dyn. **23**(1), 17–29 (1994)

Shahzad, F., Xiu, G.Y., Shahbaz, M.: Organizational culture and innovation performance in Pakistan's software industry. Technol. Soc. **51**, 66–73 (2017)

Shin, J., Mollah, M.A., Choi, J.: Sustainability and organizational performance in South Korea: the effect of digital leadership on digital culture and employees' digital capabilities. Sustainability **15**(3), 2027 (2023)

Shoss, M.K.: Job insecurity: an integrative review and agenda for future research. J. Manag. **43**(6), 1911–1939 (2017)

Soluk, J., Kammerlander, N.: Digital transformation in family-owned Mittelstand firms: a dynamic capabilities perspective. Eur. J. Inf. Syst. **30**, 1–36 (2021)

Sousa, M.J., Rocha, Á.: Skills for disruptive digital business. J. Bus. Res. **94**, 257–263 (2019)

Stettina, C.J., Heijstek, W.: Five agile factors: Helping self-management to self- reflect. Paper presented at the European Conference on Software Process Improvement (2011)

Swift, M., Lange, D.: Digital leadership in Asia-Pacific. Korn Ferry (2018)

Tabrizi, B., Lam, E., Girard, K., Irvin, V.: Digital transformation is not about technology. Harv. Bus. Rev. **13**, 1–6 (2019). https://hbr.org/2019/03/digital-transformation-is-not-about-technology
Teece, D.J.: Explicating dynamic capabilities: the nature and microfoundations of (sustainable) enterprise performance. Strateg. Manag. J. **28**(13), 1319–1350 (2007)
Teece, D.J., Pisano, G., Shuen, A.: Dynamic capabilities and strategic management. Strateg. Manag. J. **18**(7), 509–533 (1997)
Teece, D., Peteraf, M., Leih, S.: Dynamic capabilities and organizational agility: risk, uncertainty, and strategy in the innovation economy. Calif. Manag. Rev. **58**, 13–35 (2016). https://doi.org/10.1525/cmr.2016.58.4.13
Tenney, E.R., Logg, J.M., Moore, D.A.: (Too) optimistic about optimism: the belief that optimism improves performance. J. Pers. Soc. Psychol. **108**, 377–399 (2015). https://doi.org/10.1037/pspa0000018
Trenerry, B., Chng, S., Wang, Y., Suhaila, Z.S., Lim, S.S., Lu, H.Y., Oh, P.H.: Preparing workplaces for digital transformation: An integrative review and framework of multi-level factors. Front. Psychol. **12**, Article 620766 (2021). https://doi.org/10.3389/fpsyg.2021.620766
Usai, A., Orlando, B., Mazzoleni, A.: Happiness as a driver of entrepreneurial initiative and innovation capital. J. Intellect. Cap. **21**(6), 1229–1255 (2020)
Vakola, M.: The reasons behind change recipients' behavioral reactions: a longitudinal investigation. J. Manag. Psychol. **31**(1), 202–215 (2016). https://doi.org/10.1108/jmp-02-2013-0058
Vakola, M., Nikolaou, I.: Attitudes towards organizational change. Empl. Relat. **27**(2), 160–174 (2005). https://doi.org/10.1108/01425450510572685
van den Heuvel, S., Freese, C., Schalk, R., van Assen, M.: How change information influences attitudes toward change and turnover intention. Leadersh. Org. Dev. J. **38**, 398–418 (2017). https://doi.org/10.1108/lodj-03-2015-0052
Vargas-Halabí, T., Mora-Esquive, R., Siles, B.: Intrapreneurial competencies– development and validation of a measurement scale. Eur. J. Manag. Bus. Econ. **26**(1), 86–111 (2017). https://doi.org/10.1108/EJMBE-07-2017-006
Varshney, D.: Digital transformation and creation of an agile workforce: exploring company initiatives and employee attitudes. In: Turkmenoglu, M.A., Cicek, B. (eds.) Contemporary Global Issues in Human Resource Management, pp. 89–105. Emerald Publishing Limited, Leeds (2020). https://doi.org/10.1108/978-1-80043-392-220201009
Venkatesh, V., Davis, F.D.: A theoretical extension of the technology acceptance model: four longitudinal field studies. Manag. Sci. **46**(2), 186–204 (2000). https://doi.org/10.1287/mnsc.46.2.186.11926
Venus, M., Johnson, R.E., Zhang, S., Wang, X.H., Lanaj, K.: Seeing the big picture: a within-person examination of leader construal level and vision communication. J. Manag. **45**(7), 2666–2684 (2019)
Verhoef, P.C., Broekhuizen, T., Bart, Y., Bhattacharya, A., Dong, J.Q., Fabian, N., et al.: Digital transformation: a multidisciplinary reflection and research agenda. J. Bus. Res. **122**, 889–901 (2021)
Wanberg, C.R., Banas, J.T.: Predictors and outcomes of openness to change in a reorganizing workplace. J. Appl. Psychol. **85**, 132–142 (2000). https://doi.org/10.1037/0021-9010.85.1.132
Weber, E., Buettgen, M., Bartsch, S.: How to take employees on the digital transformation journey: an experimental study on complementary leadership behaviors in managing organizational change. J. Bus. Res. **143**, 225–238 (2022)
Winter, S.G.: The satisficing principle in capability learning. Strateg. Manag. J. **21**(10–11), 981–996 (2000)
Worley, C.G., Doolen, T.L.: The role of communication in implementing organizational change. J. Bus. Commun. **43**(3), 297–310 (2006)

Xu, D., Benbasat, I., Cenfetelli, R.T.: Integrating service quality with system and information quality: an empirical test in the e-service context. MIS Q. **37**(3), 777–794 (2013)

Yukl, G.: Effective leadership behavior: what we know and what questions need more attention. Acad. Manag. Perspect. **26**(4), 66–85 (2012). https://doi.org/10.5465/amp.2012.0088

Zehir, C., Karaboğa, T., Başar, D.: The transformation of human resource management and its impact on overall business performance: big data analytics and AI technologies in strategic HRM. In: Hacioglu, U. (Ed.), Digital Business Strategies in Blockchain Ecosystems, Contributions to Management Science. Springer, Cham (2020)

Zhao, H.H., Seibert, S.E., Taylor, M.S., Lee, C., Lam, W.: Not even the past: the joint influence of former leader and new leader during leader succession in the midst of organizational change. J. Appl. Psychol. **101**(12), 1730–1738 (2016). https://doi.org/10.1037/apl0000149

Bridging Bytes and Behaviors: Human-Centric Research in Emerging Digital Economies

Nicos Rossides(✉)

Department of Management and Marketing, The European University of Cyprus (& Minjiang University), 6, Diogenous Str., 2404 Nicosia, Cyprus
n.rossides@external.euc.ac.cy

Abstract. This chapter explores how international marketing research (IMR) is evolving in emerging digital economies—focusing on the Gulf States, China, and India—to balance technological innovation with a deep understanding of cultural and human dimensions. Drawing upon the Resource-Based View, Contingency Theory, and the Technology Acceptance Model, the analysis explains why market-specific research methods must adapt to local conditions. Through case studies and in-depth interviews with 15 industry professionals, the chapter demonstrates how AI-driven analytics, voice-based surveys, hybrid approaches, and integrated digital ecosystems can unlock unparalleled insights into rapidly transforming consumer behaviors. Each case reveals that while cutting-edge data tools extend research reach and granularity, cultural context, privacy considerations, and ethical safeguards are paramount. The chapter ultimately presents a forward-looking framework for IMR that synthesizes big data and human insight—offering researchers, practitioners, and global brands a roadmap for navigating complex, fast-changing markets without losing sight of the people whose decisions and motivations underpin all marketing research.

Keywords: International marketing research · Emerging digital economies · Cultural context · AI-driven analytics · Resource-Based View · Contingency Theory · Technology Acceptance Model · Ethical considerations

1 Introduction

In the streets of Mumbai, a young professional checks product reviews on her smartphone before making a purchase. In a café in Dubai, a group of friends share their experience of a new brand through local social media. In a tier-3 Chinese city, a small business owner uses a super-app to understand customer preferences and tailor his offerings. These scenes illustrate the complex consumer behavior patterns in emerging markets which are being transformed by digital technology.

Marketing researchers find themselves at a critical juncture. This digital revolution provides unprecedented tools for data collection and analysis, yet interpretation and application still require deep cultural understanding and human insight.

A. K. Zarkada (Ed.): GMA-GAMMA 2025, SPBE, pp. 54–73, 2025.
https://doi.org/10.1007/978-3-031-96930-0_4

International marketing research (IMR) is transforming beyond simple digitization toward a paradigm combining powerful technologies with human expertise (Thakur & Kushwaha, 2024).

As argued in 'The Future of Work: Managing in the Age of AI,' (Rossides, 2024), knowledge work is fundamentally changing with artificial intelligence, with marketing research one of the domains being transformed. A consequence of the huge proliferation of data is that we see a significant imbalance: over 90% of available data is unstructured—including vast amounts of unsolicited consumer opinions—yet research budgets overwhelmingly focus on solicited customer opinions, representing less than 10% of available data. This stark imbalance, a glaring disconnect between the data we have and the data we use, represents a great research opportunity—if we are able to overcome inertia and successfully embrace the new methodologies made possible by AI driven analysis, rather than resisting the transformation and potentially remaining deaf to the voices of the very customers we seek to understand.

This chapter examines IMR evolution in three key emerging markets: the Gulf States, China, and India. Each presents unique economic, technological, and societal factors influencing digital research methodology adoption. Our goal moves beyond digital transformation to focus on engaging people while efficiently integrating processes and platforms for robust decision support.

Our theoretical foundations draw from established frameworks including the Resource-Based View, Contingency Theory, and Technology Acceptance Model. These perspectives help us understand how firms develop competitive advantages through digital capabilities, adapt methodologies to diverse cultural contexts, and navigate varying technology acceptance levels across markets.

Our exploration addresses three key research questions:

1. How can researchers effectively integrate AI-driven analytics with traditional methods for deeper consumer insights in culturally diverse emerging markets?
2. What specific challenges and opportunities exist for IMR in the Gulf States, China, and India, and how do these shape effective research strategies?
3. How can researchers balance technological advancement with ethical considerations and cultural sensitivity to create truly human-centric IMR approaches?

Through case studies based on interviews with 15 industry professionals, we demonstrate how successful research approaches combine technological innovation with cultural understanding. We examine emerging technologies like virtual reality, blockchain, and synthetic data, considering their potential to revolutionize cross-cultural research while addressing ethical implications.

Our goal is to develop a holistic framework for human-centric, technologically advanced marketing research in dynamic and diverse markets. As we attempt to come to grips with increasingly technology-mediated environments, our challenge—and opportunity—is using digital tools to build bridges of empathy and insight across cultures and markets.

2 Theoretical Foundations for Digital Transformation in IMR

International marketing research transformation requires robust theoretical grounding to understand how organizations implement and benefit from digital technologies across diverse markets. Three frameworks provide complementary insights: the Resource-Based View with its dynamic capabilities extension, Contingency Theory, and the Technology Acceptance Model.

2.1 Resource-Based View and Dynamic Capabilities

The Resource-Based View significantly influences international marketing capabilities by emphasizing unique resources and competencies in achieving sustainable competitive advantage (Barney, 1991). In digital IMR, technological tools and data analytics capabilities serve as distinctive organizational resources that competitors cannot easily replicate.

According to RBV, for a resource to provide a sustainable competitive advantage, it must meet what are termed VRIN criteria: **Valuable**, meaning it enables the firm to exploit opportunities or neutralize threats; **Rare**, indicating it is possessed by few competing firms; **Inimitable**, implying it is costly or difficult to imitate; and **Non-substitutable**, meaning it lacks strategically equivalent substitutes. In the context of international marketing research, these criteria illuminate how a firm's ability to gather and analyze market intelligence can become a powerful, enduring asset (Table 1).

In the Gulf States, valuable resources include culturally-calibrated AI models that accurately interpret the nuanced ways religious and social norms influence online expression and purchasing behavior. Foreign firms often struggle to develop these resources independently, making local partnerships particularly valuable. As a consequence, the firms that establish these partnerships gain significant advantages in consumer understanding that competitors cannot easily match.

For China, valuable resources extend beyond technical capabilities to include government relationships and regulatory knowledge essential for research operation. The closed nature of China's digital ecosystem means that access itself becomes a VRIN resource. Indeed, companies that have established data-sharing agreements with platforms like WeChat or Alibaba possess rare and valuable assets that provide sustainable competitive advantage in consumer insight generation. (Horstmann, 2021).

In India, the most valuable resources often involve methodological flexibility and multi-modal approaches that bridge the pronounced urban-rural and socioeconomic divides. Organizations that develop voice-based technologies capable of processing diverse regional dialects gain access to consumer segments largely invisible to traditional research methods. That makes a real difference, transforming previously inaccessible market insights into a rare, inimitable, and non-substitutable competitive advantage, effectively unlocking the potential of a vast and diverse market.

Dynamic capabilities extend RBV by emphasizing how firms adapt and reconfigure resources to maintain advantage in changing environments (Teece et al., 1997).

Table 1. RBV Applications in International Marketing Research

Resource Type	Examples in IMR	Market-Specific Applications	Competitive Advantage
Technological	Proprietary analytics algorithms; Custom research platforms	Gulf States: Arabic sentiment analysis tools; China: WeChat mini-program integration; India: Voice-based multilingual surveys	Enhanced data processing; Unique insights; Access to underrepresented populations
Human	Cross-cultural expertise; Analytical specialists	Gulf States: Understanding of religious influences on consumer behavior; China: Knowledge of digital ecosystem regulations; India: Linguistic expertise across regional dialects	Cultural nuance detection; Regulatory navigation; Comprehensive representation
Informational	Proprietary databases; Exclusive market access	Gulf States: High-income consumer panels; China: Super app usage data; India: Rural consumer insights	Exclusive segment understanding; Integrated behavioral insights; Market gap identification
Organizational	Integrated research workflows; Knowledge management systems	Gulf States: Hybrid digital-traditional methodologies; China: Local partnership networks; India: Multi-tiered research approaches	Methodological flexibility; Enhanced market access; Comprehensive demographic coverage

While RBV focuses on static resource possession, dynamic capabilities capture the processes through which firms evolve their capabilities in response to market changes. These manifest differently across emerging markets:

China:
Within China's dynamic technological environment, successful research organizations excel at sensing and adapting to changes, such as new *super app* features or government regulations. When WeChat launched mini-programs—small, in-app applications that provide specific services without requiring separate downloads—research firms possessing strong dynamic capabilities swiftly developed innovative data collection methodologies, securing a first-mover advantage in generating valuable insights.

Gulf States:
In the Gulf States, effective dynamic capabilities hinge on cultural adaptation. As social media usage evolves among younger demographics, research organizations must continuously recalibrate their approaches and privacy considerations. Firms that successfully navigate this changing environment maintain legitimacy and access; this then helps them gain a crucial edge in capturing evolving cultural nuances.

India:
For India, dynamic capabilities center on technological bridges between traditional and digital methodologies. When demonetization rapidly accelerated digital payment adoption, adaptive research firms quickly developed hybrid methodologies. These firms tracked shifting financial behaviors across demographics, delivering timely, comprehensive insights to clients that neither pure digital nor traditional methods could provide.

Dynamic Capabilities and Resilience:
Research indicates dynamic capabilities mediate the relationship between RBV and organizational resilience during market disruptions (You et al., 2023). This relationship proved particularly evident during COVID-19, when research organizations with strong dynamic capabilities rapidly pivoted from face-to-face methodologies to digital approaches without sacrificing insight quality or representation. Indeed, at the heart of dynamic capabilities lies the ability to sense, seize, and reconfigure resources in response to dynamic environments, ensuring continued value creation and competitive advantage.

2.2 Contingency Theory in Cross-Cultural Research

Contingency theory provides a framework for understanding how research methodologies should adapt to different market contexts. The theory posits organizational success depends on aligning internal characteristics with external circumstances—there is no universal approach, and optimal strategy depends on specific environments (Ayman et al., 1995). Essentially, if certain external circumstances exist, then a specific research approach is optimal.

For International Marketing Research (IMR), Contingency Theory reveals why methodologies effective in one national context fail in others. Market-specific analysis through this lens identifies critical circumstances that research approaches must address:

In China, the integrated nature of digital platforms creates a unique circumstance unlike any Western digital environment. Research methods designed for fragmented digital ecosystems fundamentally misalign with this environment. Successful approaches must be redesigned specifically for the *super app* ecosystem, where consumer journeys span seamlessly across previously distinct domains like social media, e-commerce, financial services, and entertainment within a single platform.

For Gulf States, critical contingencies include the role of family decision-making in purchasing behavior, religious influences on product perception, and distinct gender-based differences in digital platform usage. Research methodologies must adapt to these contingencies through appropriate segmentation, culturally-sensitive questioning approaches, and careful consideration of how, when, and where consumers are approached.

In India, the contingencies requiring methodological alignment vary dramatically between urban and rural environments, creating essentially different research contexts within the same national market. Urban areas with high smartphone penetration and digital literacy allow sophisticated digital methodologies, while rural contexts with limited connectivity and varied literacy levels necessitate traditional or voice-based approaches. In response, successful research organizations develop bifurcated methodologies that align with these divergent contingencies rather than forcing standardized approaches.

Contingency Theory emphasizes flexibility over standardization—the opposite of traditional global research approaches that prize methodological consistency. This theoretical perspective supports the development of market-specific research frameworks calibrated to local conditions rather than globally standardized approaches that miss critical cultural and technological nuances. Eschewing universal templates, effective international marketing research relies on adapting to the unique interplay of environmental factors in each market—from a 'one-size-fits-all' mentality to a contextual understanding of research design. It acknowledges that what constitutes 'rigor' or 'validity' can vary significantly across cultures and technological environments, a concept that may challenge those rooted in positivist research paradigms. Consequently, according to this theory, research organizations must cultivate a dynamic capability to not only gather data, but also to interpret it through a culturally nuanced lens; this, to ensure that insights remain relevant and actionable in each specific market.

2.3 Technology Acceptance Model and Digital Adoption

The Technology Acceptance Model provides insights into how new technologies, including research tools and platforms, are adopted across different markets (Davis, 1989). TAM identifies two key factors influencing user attitude and behavioral intention to use new technology: Perceived Usefulness (PU) and Perceived Ease-of- Use (PEOU).

This model has profound implications for research methodology selection and implementation across diverse markets with varying technological acceptance patterns:

In the Gulf States, high smartphone penetration suggests favorable conditions for mobile research. But TAM analysis reveals that privacy concerns significantly impact perceived usefulness of research applications, particularly among certain demographic segments. Successful research approaches address these concerns explicitly through culturally appropriate privacy assurances and transparent data handling explanations.

For China, perceived usefulness often relates to social validation and integration with existing digital habits. Research methodologies embedded within popular super apps achieve higher participation rates than standalone research applications, regardless of technical sophistication. This represents a direct application of TAM principles—aligning research tools with established digital behaviors reduces adoption friction.

In India, perceived ease-of-use varies dramatically across demographic segments. Urban professionals readily adopt sophisticated research applications, while rural and older populations may struggle with text-based interfaces regardless of simplified design.

Voice-based research interfaces succeed precisely because they align with TAM principles—they minimize the technical skills required while maximizing accessibility across literacy levels.

TAM also explains participation biases in online research panels. Segments with greater technology comfort are systematically overrepresented in digital research, creating embedded biases in resulting insights. Understanding these dynamics through TAM allows researchers to identify representation gaps and develop corrective methodologies.

The integration of these three theoretical perspectives—RBV and dynamic capabilities, Contingency Theory, and TAM—provides a comprehensive framework for understanding the complexities of international marketing research in emerging digital economies. Together, they explain why certain research approaches succeed in specific markets while failing in others, and they guide the development of methodologically sound, culturally sensitive research strategies adapted to each unique market context.

With this theoretical foundation established, we can now examine how international marketing research has evolved and continues to transform in practice.

3 The Evolving Environment of International Marketing Research

International marketing research has transformed dramatically, driven by technological advancement, changing consumer behaviors, and market globalization. This evolution has revolutionized data collection, analysis, and interpretation practices across borders.

Traditional methods like face-to-face interviews dominated the 1970s and 1980s, demanding substantial resources while limiting geographical reach. The 1990s internet adoption enabled cost-effective online surveys, though sample representativeness and uneven internet penetration created new challenges (Craig & Douglas, 2001).

A critical issue in current practice is data utilization imbalance. While unstructured data comprises over 90% of available information, research budgets predominantly target solicited customer opinions. Companies relying exclusively on surveys might miss valuable unsolicited feedback from online communities, creating significant insight gaps.

AI's transformative power extends beyond basic automation as seen in Table 2:

Table 2. AI Applications in International Marketing Research

Application	Function	Strategic Benefit
Real-Time Data Processing	Generates continuously updated insights	Enables proactive response to market shifts
Holistic Consumer Analysis	Integrates diverse data sources	Reveals connections between behaviors and cultural trends
Advanced Analytics	Identifies correlations and suggests causality	Supports predictive decision-making
Enhanced Decision Support	Simulates business scenarios	Delivers personalized insights across languages

AI-powered analysis combined with traditional techniques offers comprehensive understanding beyond what either approach alone provides. AI technologies have revolutionized consumer insight discovery, advertising creation, and impact evaluation—enabling process automation and deeper customer understanding (Chen, 2024).

Companies leading this transformation include Epsilon (extensive data points), Quantcast (data-driven personas), and NextRoll (automated marketing campaigns). These platforms demonstrate how AI unlocks valuable behavioral insights—which in turn leads to a deeper understanding of consumers and an improved ability to anticipate market changes (Kotha, 2024).

4 Market-Specific Challenges and Opportunities

Gulf States: Navigating Cultural Complexity in a Mobile-First Environment
The Gulf States present unique research opportunities due to their exceptionally high smartphone penetration rates, often exceeding 97% in 2024 (Statista, 2025). This mobile-first environment provides a rich environment for digital research—although the region's cultural nuances add significant complexity to these efforts (Jahan et al, 2022).

Personal relationships remain central to business transactions in the Gulf. This cultural emphasis necessitates hybrid research approaches that combine digital data collection with face-to-face methods, enabling researchers to build trust and rapport. This, though, requires navigating strict data protection laws while remaining sensitive to cultural norms around privacy.

AI tools analyzing Arabic social media conversations offer nuanced insights into consumer sentiment that traditional methods often miss. Having said that, local researchers must be mindful of cultural norms governing public discourse and anticipate social desirability bias (Salem & Mourtada, 2012).

Ultimately, successful research in the Gulf requires cultural competence in addition to technical expertise. Understanding family dynamics, religious considerations, and gender norms is essential for accurate data interpretation and ensuring research activities respect local values and norms.

China: Super Apps and Integrated Digital Ecosystems
China's vast population and unique digital ecosystem create distinctive research conditions that demand a nuanced understanding. Platforms like WeChat and Alibaba have established a digital environment fundamentally different from those in the West. These *super apps* integrate a vast array of services, from social media and messaging to e-commerce, payments, and transportation—offering unprecedented access to consumer data (Keane & Chen, 2019).

This closed ecosystem, however, presents barriers for international researchers seeking insights. Platform regulations, government policies, and technical integration challenges complicate data collection, especially if an international agency has not established strong local partnerships.

Furthermore, technological advancements in China occur at a remarkable pace. New innovations are rapidly adopted by the population, which then forces continuous adaptation of research methodologies. For instance, the widespread use of mobile payments and social commerce blurs traditional boundaries between data types. This creates rich opportunities for holistic consumer understanding while simultaneously raising significant privacy questions (Zhang et al., 2021).

It therefore follows that research methods designed for China's integrated ecosystem differ substantially from those effective in fragmented digital environments. Traditional survey-based approaches, for example, may miss crucial interconnected insights available within these single app ecosystems. It therefore becomes necessary for researchers to adopt innovative methodologies that can capture the full spectrum of consumer behavior within these dynamic and interconnected platforms. Very few, if any, international contexts are characterized by such dynamic and interconnected platforms—at least not to the same degree.

India: Addressing the Digital Divide and Linguistic Diversity

In stark contrast to China's unified digital ecosystem, India presents a research environment characterized by stark contrasts and a variety of distinct challenges. The country's vast population and significant digital divide between urban and rural areas necessitate a multi-pronged approach. While cutting-edge digital methodologies thrive in urban centers, traditional approaches remain essential for reaching rural populations with limited internet access and digital literacy (Kumar & Gupta, 2021).

Linguistic diversity adds another layer of complexity. With 22 officially recognized languages and hundreds of dialects, capturing nuanced insights across India's linguistic spectrum poses substantial challenges. While AI-powered translation tools have advanced significantly, accurately rendering subtleties across numerous languages remains very challenging.

Innovative researchers have developed voice-based surveys and visual interfaces to overcome literacy barriers and gather data from previously underrepresented populations. These methodologies bridge gaps that traditional text-based research would miss entirely.

Furthermore, India's research context features significant urban-rural differences in technology adoption patterns. The Technology Acceptance Model explains these variations, highlighting different perceptions of usefulness and ease-of-use across demographic groups.

Across these diverse markets, a common thread emerges: successful research balances technological innovation with cultural sensitivity and human insight. As Rossides (2024) argues in *The Future of Work*, AI's capabilities complement rather than replace human expertise in knowledge-based domains like marketing research, shifting researcher roles toward interpretation, strategy formulation, and ethical oversight.

The ethical implications of these technological advancements cannot be overstated. As researchers gather increasingly detailed consumer behavior data, questions of privacy, consent, and data ownership become paramount. In emerging markets with evolving regulatory frameworks, researchers bear particular responsibility for ensuring ethical practices that respect local norms (Markham, & Buchanan, 2015).

To illustrate how these market-specific challenges are addressed in practice, we now turn to case studies from each region based on in-depth interviews with industry professionals.

Ensuring Data Quality in Emerging Markets

In international marketing research, the quality of data—its accuracy, reliability, and representativeness—can make or break the insights derived from it. Yet, in emerging markets, where technological infrastructure, cultural diversity, and regulatory environments vary widely, ensuring data quality is no simple task. As one industry professional noted during our interviews, *"In these markets, data isn't just numbers—it's a reflection of complex, often fragmented realities."* This complexity demands a nuanced approach to data collection and validation.

One recurring theme from our interviews was the challenge of accuracy in data collection. In regions with limited digital penetration, such as rural India, traditional methods like face-to-face interviews often yield more reliable results than online surveys. However, as one researcher pointed out, *"Even traditional methods have their limits. You need to cross-verify findings through multiple approaches to ensure you're capturing the truth."* For example, combining voice-based surveys with in-person interviews can help validate responses, especially in areas where literacy levels vary.

Representativeness is another critical concern. Emerging markets are often characterized by stark contrasts—between urban and rural areas, between tech-savvy millennials and older generations, or between speakers of different languages and dialects. A marketing executive we spoke to emphasized, *"If your sample doesn't reflect these divides, your insights will be skewed."* Stratified sampling, which ensures adequate representation from all segments, is essential. In China, for instance, researchers often design studies to capture both urban consumers in tier-1 cities and rural populations in tier-3 cities, ensuring a balanced view of the market.

The rise of unstructured data—from social media posts to online reviews—has added both opportunities and challenges. While this data offers rich, real-time insights, it also introduces noise and irrelevant information. AI tools can help filter and analyze this data, but as one data scientist cautioned, *"AI is only as good as the human oversight behind it. Without cultural context, algorithms can miss the subtleties that matter most."* In the Gulf States, for example, AI-driven sentiment analysis of Arabic social media data is often complemented by manual reviews to ensure culturally specific expressions are accurately interpreted.

Finally, cultural sensitivity plays a pivotal role in ensuring data quality. Researchers must design data collection methods and questions that resonate with local norms and values. As one interviewee put it, *"What works in one market can backfire in another if you don't account for cultural differences."* In the Gulf States, for instance, surveys are often adapted to reflect the influence of family and religious values on consumer behavior, ensuring that the data captures the true drivers of decision-making.

These insights from industry professionals underscore a key point: ensuring data quality in emerging markets is both a technical challenge and, in many ways, a strategic imperative. The way forward is to combine rigorous methodologies with cultural understanding. By doing so we can unlock insights that are both accurate and actionable—bridging the gap between data and real-world decision-making.

5 Case Studies in Emerging Digital Economies

Our research includes interviews with 15 industry professionals across the Gulf States, China, and India. These cases illustrate practical applications of innovative research approaches.

Our case studies represent actionable frameworks rather than isolated examples. Each demonstrates strategic approaches that organizations can adapt to their specific needs in these complex markets. We provide concrete recommendations derived from both successful implementations and learning opportunities from these real-world applications.

Gulf States: Balancing Technology and Tradition

A multinational consumer goods company investigating sustainability attitudes among young Gulf consumers employed a hybrid approach combining AI-driven social media analysis with focus groups.

Their AI analysis using Arabic-tailored natural language processing identified growing sustainability interest among young consumers. The team supplemented this with in-person focus groups to gain context.

"Focus groups revealed that sustainability interest wasn't simply environmental concern. It stemmed from desires to align with global trends while reinterpreting traditional resource conservation values," explained their marketing executive.

This case demonstrates how technology enhances human insight and the importance of culturally-adapted methodologies. For companies entering Gulf markets, this approach offers a strategic template: begin with AI analysis of social conversation, but verify and contextualize findings through traditional qualitative research that respects local cultural norms.

China: Leveraging Integrated Digital Platforms for Consumer Insights

A global automotive manufacturer studied electric vehicle preferences by leveraging China's unique digital ecosystem. The company analyzed data from WeChat's integrated services spanning social media, e-commerce, and payment systems—an approach impossible in Western markets with fragmented digital environments.

"We developed AI models analyzing data across multiple touchpoints: social discussions, search behaviors, content consumption, and purchasing patterns. The integration of these data streams provided unprecedented insight into the complete consumer journey," explained their data scientist.

However, this data wealth presented interpretation challenges. "We discovered significant discrepancies between expressed preferences on social platforms and actual purchasing behaviors tracked through e-commerce and payment data. This prompted us to establish dedicated online communities for deeper qualitative exploration," noted their senior researcher.

For companies seeking to conduct research in China's unique digital ecosystem, three specific strategies emerge: First, develop dedicated analytics tools that integrate data across WeChat's various functions rather than treating channels separately. Second, establish formal partnerships with local technology firms for essential platform access. Third, implement continuous monitoring systems for regulatory changes that might affect data collection or analysis capabilities. These approaches help international firms navigate both technological and regulatory complexities.

The combined methodology revealed nuanced insights that challenged established assumptions. While social media analysis showed high interest in long-range electric vehicles, community discussions and behavioral data revealed urban Chinese consumers prioritized charging infrastructure density for short city trips— directly contradicting Western assumptions about range anxiety as the primary adoption barrier.

The research team also uncovered significant regional variations. "Purchase motivations in tier-1 cities centered on avoiding license plate restrictions for combustion engines, while tier-2 city consumers emphasized status signaling through technology adoption. These distinctions required fundamentally different marketing approaches," the senior researcher added.

This case illustrates three critical lessons for China market research: First, integrated ecosystems demand integrated methodologies rather than siloed channel analysis. Second, expressed preferences and actual behaviors often diverge significantly, requiring triangulation across methodologies. Third, local expertise remains essential for navigating both technological platforms and cultural interpretation, making partnerships with Chinese research organizations particularly valuable.

India: Innovative Approaches to Overcome Barriers

A fintech company investigating financial behaviors in tier-2 and tier-3 cities developed innovative research approaches to address India's unique challenges.

"We created a voice-based survey tool that administered questionnaires in multiple Indian languages and dialects. This proved essential for reaching respondents who would struggle with text-based surveys," explained their marketing executive.

The voice responses underwent analysis using AI-powered natural language processing capable of understanding and categorizing responses across different languages. "This technology allowed efficient processing of large qualitative data volumes, identifying patterns that human analysts might miss," noted their data scientist.

Organizations entering India should implement a three-tier research strategy: First, deploy standard digital methodologies in urban areas with high connectivity.

Second, develop voice-based tools with regional language capabilities for semi-urban regions. Third, maintain traditional face-to-face approaches for rural areas with minimal digital penetration. This stratified approach ensures representative insights across India's diverse technological landscape. Particularly effective are visual research tools that use universal imagery to transcend both literacy and linguistic barriers.

The research team recognized the limitations of this technology-driven approach. "We complemented voice surveys with in-person interviews conducted by local researchers. These interviews provided context for trends identified by AI analysis and uncovered cultural nuances the technology missed," emphasized a senior researcher.

While AI analysis identified increased digital payment adoption, in-person interviews revealed this trend stemmed not just from convenience but also from changing social norms around family money management.

This case demonstrates how technology can overcome scale and linguistic diversity challenges. It also highlights human insight's continued importance in interpreting and contextualizing data, especially in culturally diverse markets like India.

6 Evolving Skill Sets for Modern Marketing Researchers

Today's international marketing researchers face dramatically different capability requirements than their predecessors. The digital transformation of research demands continuous adaptation and skills evolution across multiple domains.

AI Collaboration and Augmentation

The ability to effectively partner with AI systems represents perhaps the most critical emerging skill. Researchers must become proficient in prompt engineering—crafting precise instructions that generate optimal AI outputs tailored to specific research questions. As one data scientist noted, "The difference between mediocre and exceptional insights often lies in how researchers formulate their AI interactions."

This collaboration requires understanding AI strengths and limitations across different cultural contexts. Researchers must recognize when algorithms might miss cultural nuances and when they might extract patterns invisible to human analysis. The most effective professionals develop what may be termed "complementary specialization"—focusing their efforts on tasks requiring astute human judgment while delegating pattern recognition and processing to AI systems (Rossides, 2024).

AI augmentation skills extend to research design and methodology selection. Forward-thinking researchers use AI systems to simulate different methodological approaches before implementation, testing potential outcomes and identifying cultural biases in advance. This capability dramatically improves research efficiency while reducing implementation risks.

Agility and Continuous Upskilling

The accelerating pace of technological change requires perpetual learning habits. Successful researchers establish systematic approaches to skills development rather than episodic training. As one senior executive explained, "We expect researchers to dedicate 20% of their time to learning new methodologies, technologies, and analytical approaches. Yesterday's expertise becomes today's obsolescence without this commitment."

Research organizations increasingly implement formal upskilling frameworks that combine structured learning paths with applied projects. These programs emphasize practical application of emerging technologies like synthetic data generation, voice-based research, and augmented reality testing environments.

Technical agility must be matched by methodological flexibility. Researchers need to quickly adapt approaches when market conditions change or technologies evolve. During China's rapid shift to mini-program dominance, researchers who quickly mastered these environments gained significant advantages over those who maintained traditional methodologies.

Distributed Collaboration Models

The modern researcher operates within fluid team structures that span geographies, organizations, and employment models. The ability to collaborate effectively with freelance specialists, AI systems, and cross-functional stakeholders simultaneously has become essential for complex international projects.

"Traditional research departments are giving way to networked expertise models," observed the research director at a multinational consumer goods company. "Our China market study combined our core team with freelance platform specialists, local academic partners, and AI-driven analytics systems. Managing these diverse contributors requires sophisticated coordination skills."

Remote collaboration technologies have eliminated geographical constraints while introducing new challenges. Researchers must develop skills for virtual team management, asynchronous collaboration, and digital knowledge sharing. These capabilities prove particularly important for international projects spanning multiple time zones and cultural contexts.

Distributed research models create opportunities for specialized expertise integration that was previously impossible. A healthcare research project might seamlessly combine AI-driven social listening, freelance ethnographers in rural communities, and academic experts providing theoretical frameworks—all coordinated by a core research team using collaborative platforms.

Synthesis and Integration

Perhaps the most valuable skill in this evolving landscape is the ability to synthesize insights across diverse sources and methodologies. Researchers must integrate AI-generated patterns, human observations, quantitative measurements, and cultural contexts into coherent strategic narratives that drive decision-making.

This integration capability distinguishes exceptional researchers from technological operators. While AI systems excel at finding patterns, and specialized collaborators provide deep expertise in narrow domains, the researcher's unique value lies in connecting these elements into meaningful insights that address specific business challenges.

The modern marketing researcher serves as both orchestrator and integrator—directing a complex ecosystem of technologies, methodologies, and collaborators while synthesizing their contributions into actionable intelligence. This evolving role demands continuous adaptation, technological fluency, and collaborative expertise well beyond traditional research skills.

7 Ethical Imperatives in Digital IMR

Ethical considerations grow increasingly complex as research embraces digital technologies. Privacy protection varies dramatically across regions, creating challenges for cross-border projects. The European Union's GDPR establishes stringent standards, while other regions operate under less comprehensive frameworks.

AI models trained primarily on Western data often perform poorly in different cultural contexts. Addressing algorithmic bias requires diverse, representative training datasets and continuous monitoring. Market-specific models may prove necessary in distinctive cultural environments.

Emerging technologies create additional ethical considerations. Virtual reality can produce unintended psychological effects. Blockchain's immutable records raise consent questions. Synthetic data generation creates complex representation issues (Table 3).

Table 3. Ethical Challenges in Digital IMR

Challenge	Description	Mitigation Strategy
Data Privacy	Varied regulatory environments	Adopt highest protection standards regardless of local requirements
Algorithmic Bias	Cultural misinterpretation of sentiment	Develop market-specific training data and models
Informed Consent	Complexity of data usage	Create simplified, culturally appropriate consent processes
Data Sovereignty	Cross-border data transfer restrictions	Implement localized storage and processing
Representation	Exclusion of digitally disadvantaged groups	Combine digital with traditional research methods

Market-specific ethical frameworks must account for varying regulatory environments. In China, data localization requirements necessitate careful server location planning and clear data handling protocols. Gulf States research must navigate religious and cultural sensitivities, particularly regarding family and gender topics. Indian research ethics require special attention to vulnerable populations with limited digital literacy who may not fully comprehend digital consent processes.

Establishing regional ethics review boards with local expertise represents best practice for international organizations. These boards should include local cultural experts alongside data specialists and legal advisors. Regular ethics audits should examine not only regulatory compliance but cultural appropriateness and representation adequacy across all demographic segments.

Ethical oversight structures provide essential frameworks for maintaining high standards. Ethics committees, codes of conduct, and regular compliance audits create accountability systems protecting participant interests while enhancing research quality.

8 Future Horizons in IMR

The future of international marketing research will be shaped by emerging technologies adapted to each market's unique characteristics, consumer behaviors, and regulatory environments.

Immersive Technologies

Virtual and augmented reality offer transformative research potential when calibrated to market conditions. In China, companies leverage VR for behavioral observation in simulated shopping environments, providing insights impossible through traditional methodologies. Alibaba's VR shopping experiences, for instance, reveal moment-by-moment decision patterns without physical store investments.

Middle Eastern applications require careful cultural adaptation. VR environments must respect modesty norms and appropriate social interaction patterns. One Gulf- based researcher noted, "Virtual environments that violate cultural expectations create artificial behaviors that fundamentally undermine research validity."

India's implementation challenges center on infrastructure constraints. Mobile-based AR solutions with minimal hardware requirements prove more effective than equipment-intensive VR systems. Visual overlays on familiar environments generate higher participation rates across socioeconomic segments than fully immersive alternatives.

Synthetic Data and Its Expanding Role

Synthetic data is rapidly reshaping international marketing research, offering a powerful solution to address privacy concerns, mitigate data limitations, and enable innovative research methodologies. This technology is particularly valuable in markets with stringent data protection regulations or cultural sensitivities.

However, its benefits extend far beyond privacy protection, influencing how data-driven insights are generated, shared, and applied in dynamic market environments.

Yet, as a Global Marketing Insights Director at a leading FMCG company in the Gulf noted, synthetic data is not a magic bullet. Its effectiveness is fundamentally dependent on the quality and availability of the real-world data it seeks to replicate. Without a strong foundation of actual data, the ability of synthetic data to generate meaningful and reliable insights is significantly diminished.

For instance, in a recent tracking study, one of the director's research suppliers faced an unexpected sampling issue, failing to capture responses from a crucial sub-segment within a broader age group (e.g., young adults aged 18–24 with distinct lifestyle attributes). While the study met overall demographic quotas, the complete absence of this particular sub-group posed a major gap. Despite having access to longitudinal data, reconstructing the missing segment using synthetic data proved difficult. This raised concerns about synthetic data's reliability when compensating for entirely missing data points, particularly in cases where prior datasets lacked sufficient depth or granularity. As the director put it, "Frankly, I was disappointed.

We were counting on synthetic data to fill this critical gap, and it just couldn't deliver. That left a very sour taste as I was really expecting a lot more. At the end of the day it taught us this: without solid foundational data, synthetic data is essentially useless."

This case highlights an important nuance: while synthetic data excels at augmenting existing datasets and correcting minor imbalances, it struggles when trying to create entirely absent segments from scratch. Its success hinges on the robustness of historical data for that specific group. If past data collection efforts were incomplete or flawed, synthetic data may fail to generate realistic and actionable insights.

Blockchain for Transparency and Consent

As marketing research evolves, blockchain technology is emerging as a powerful tool for enhancing data integrity, participant control, and regulatory compliance. By leveraging decentralized ledgers and smart contracts, blockchain enables secure, verifiable, and automated consent mechanisms that address longstanding concerns about data misuse and transparency. Unlike traditional research models, where participants must trust centralized entities to handle their data ethically, blockchain allows individuals to set precise permissions, track usage, and revoke access as needed, ensuring greater control over personal information.

This is particularly significant in markets where institutional trust remains low, or regulatory environments are complex. In China, where strict data-sharing laws limit cross-platform research, blockchain creates opportunities for decentralized identity management, enabling compliant yet rich data aggregation. In the Gulf, where cultural sensitivities shape data collection norms, blockchain offers a tamper-proof system for consent management, allowing brands to engage consumers more transparently. Meanwhile, in India, where consumer data protection regulations are rapidly evolving, blockchain can support ethical data exchanges, helping research firms align with both local policies and global best practices.

Beyond privacy protection, blockchain also enables new models of participant engagement and data monetization. Researchers can tokenize data access, allowing individuals to participate in studies with defined compensation mechanisms while ensuring ethical data usage. In markets where concerns over misuse of consumer insights have previously deterred participation—such as financial services research in the Gulf or sentiment analysis in politically sensitive sectors—this level of transparency can drive significantly higher response rates.

While blockchain's potential to redefine trust and accountability in marketing research is clear, adoption remains in its early stages. The challenge lies in scaling implementation—balancing cost, technical expertise, and regulatory alignment.

Yet, for research firms willing to invest in blockchain-verified consent models, the long-term benefits are substantial: higher-quality data, stronger consumer relationships, and a more ethical, future-proof approach to insights generation in some of the world's fastest-growing markets.

AI Beyond Automation and Analysis: The Research Partner

The most profound transformation in international marketing research is AI's evolution from an automation tool to an analytical powerhouse and, ultimately, a research partner. Initially deployed to handle time-consuming, repetitive tasks—such as data cleaning, categorization, and survey administration—AI has now progressed to higher-order functions, including pattern recognition, predictive modeling, and cultural adaptation.

This shift is especially evident in Natural Language Processing (NLP), where AI has moved beyond simple translation and transcription to capturing cultural nuance, sentiment, and implicit meaning. In multilingual markets like China, India, and the Gulf, AI-driven sentiment analysis now detects subtle emotional variations and contextual shifts, bridging linguistic and social gaps in ways human analysts alone could not.

Beyond analysis, AI systems are now co-creating research methodologies, drawing from past studies to suggest culturally relevant questions, demographic weightings, and optimal survey structures. By integrating real-time data insights with historical research, AI enhances research design itself, making methodologies more adaptive, inclusive, and precise.

As one research director put it, *"AI doesn't replace human researchers; it transforms their capabilities—just as telescopes transformed astronomy. We see patterns and possibilities invisible to previous generations."*

Proposed Human-Centric IMR Framework
Bringing together the theoretical perspectives (Resource-Based View, Contingency Theory, and the Technology Acceptance Model) with the insights gleaned from our case studies and forward-looking trends, Fig. 1 illustrates an integrated Human-Centric IMR Framework. It shows how technological capabilities (AI, blockchain, synthetic data) should be balanced with cultural sensitivity, ethical oversight, and ongoing researcher involvement to yield actionable insights in emerging markets.

In this model, each component supports the others: technology automates and amplifies data collection and analysis; human researchers contribute nuanced interpretation and empathy; and robust ethical guidelines keep both the technology and human components aligned with local norms and regulations. By viewing IMR through this holistic lens, practitioners can design research programs that not only embrace digital innovations but also respect and reflect the cultural realities of their audiences.

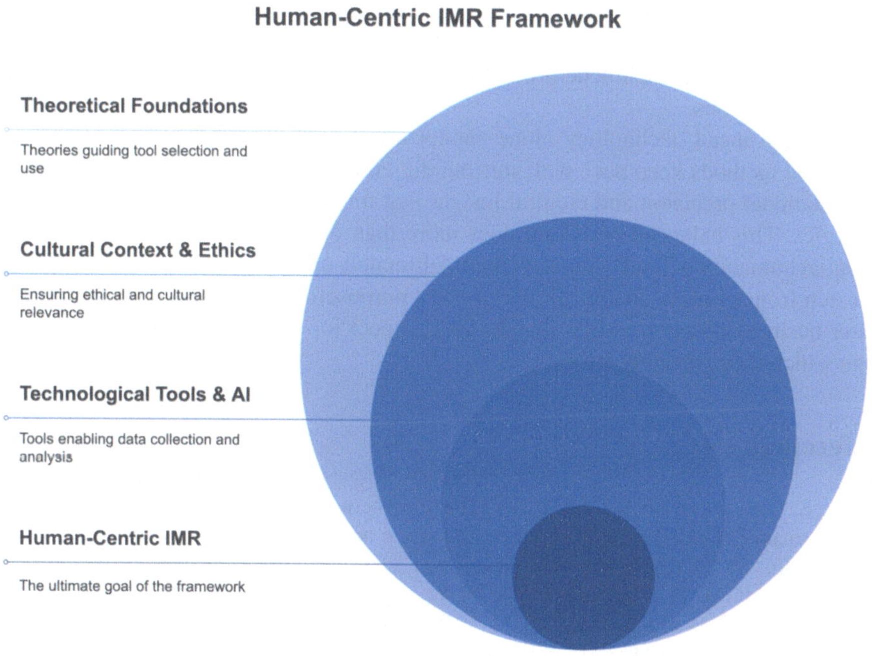

Fig. 1. .

9 Conclusion: A New Paradigm for International Marketing Research

International marketing research today stands at a defining crossroads, shaped by the interplay of advanced technologies, evolving ethical considerations, and the imperative to capture consumers' lived realities across diverse markets. The integrated, human-centric approach presented above underscores the need to harness digital innovations without losing sight of the people whose aspirations and experiences drive all commercial activity.

Our exploration of research environments in the Gulf States, China, and India underscores the challenges as well as opportunities of this transformation. The Resource-Based View explains how firms can leverage cutting-edge research tools to build sustainable competitive advantages. The Technology Acceptance Model sheds light on adoption barriers and facilitators across diverse populations.

Contingency Theory reminds us that research methodologies must be contextually adaptive, rather than relying on standardized, one-size-fits-all approaches.

Yet despite these innovations, the fundamental mission of international marketing research remains unchanged: understanding human behavior and motivations. In an era awash with data, the true challenge—and opportunity—lies in bridging bytes and behaviors. While artificial intelligence can sift through volumes of information and reveal patterns that humans alone might miss, it remains the researcher's responsibility to interpret these signals ethically and contextually. Indeed, the synergy of AI-driven analytics and cultural nuance can illuminate new dimensions of consumer behavior; yet only thoughtful, empathetic professionals can ensure these insights reflect genuine human motivations.

Looking ahead, technology alone cannot uncover the full story—nor can purely traditional methods keep pace with shifting digital realities. It is the artful synthesis of computational precision and cultural insight that transforms raw data into meaningful strategies. This balanced paradigm does more than optimize processes: it honors the essential humanity behind every data point, ultimately elevating international marketing research from a purely analytical exercise to a purposeful enterprise—one that informs better business decisions while maintaining respect for the communities and cultures from which those insights arise.

References

Afriyie, S., Duo, J., Appiah, K., Musah, A.I.: The nexus between types of innovation and marketing performance of small medium-sized enterprises in an emerging economy. Int. Rev. Manag. Mark. **8**(6), 78–92 (2018)

Alonazi, M.: MGAUM : a new framework for the mobile government service adoption in Saudi Arabia. University of Sussex (2019). https://hdl.handle.net/10779/uos.23473433.v1

Ayman, R., Chemers, M.M., Fiedler, F.: The contingency model of leadership effectiveness: Its levels of analysis. Leadersh. Quart. **6**(2), 147–167 (1995)

Barney, J.: Firm resources and sustained competitive advantage. J. Manag. **17**(1), 99–120 (1991)

Bertram, A.: Global learning in the 21st century. In: Barkatsas, T., Bertram, A. (eds.) Global Learning in the 21st Century. Sense Publishers (2016)

Chen, M.: Role of artificial intelligence in marketing automation in China. Int. J. Strateg. Mark. Pract. **6**(2), 25–35 (2024)

Coppola, C., Vollero, A., Siano, A.: Developing dynamic capabilities for the circular economy in the textile and clothing industry in Italy: a natural-resource-based view. Bus. Strateg. Environ. **32**(7), 4798–4820 (2023)

Craig, C.S., Douglas, S.P.: Conducting international marketing research in the twenty-first century. Int. Mark. Rev. **18**(1), 80–90 (2001)

Davis, F.D.: Perceived usefulness, perceived ease of use, and user acceptance of information technology. MIS Q. **13**(3), 319–340 (1989)

Erevelles, S., Fukawa, N., Swayne, L.: Big data, consumer analytics, and the transformation of marketing. J. Bus. Res. **69**(2), 897–904 (2016)

Gupta, S., Malhotra, N.: Marketing innovation: a resource-based view of international and local firms. Mark. Intell. Plan. **31**(2), 111–126 (2013)

Horstmann, N.: Digital marketing in China: how B2B companies can successfully expand into the chinese market. In: Seebacher, U.G. (eds.) B2B Marketing. Management for Professionals. Springer, Cham (2021)

Jahan, S., Bhaumik, A., Abid, S., Tiwari, K.: Consumer behaviour towards mobile marketing in UAE restaurants: application of technology acceptance model. In: Fuzzy Systems and Data Mining VIII (2022). https://doi.org/10.3233/FAIA220390

Jarek, K., Mazurek, G.: Marketing and artificial intelligence. Cent. Eur. Bus. Rev. **8**(2), 46–55 (2019)

Keane, M., Chen, Y.: Entrepreneurial solutionism, characteristic cultural industries, and the Chinese dream. Int. J. Cult. Policy **25**(6), 743–755 (2019)

Kotha, S.: The convergence of artificial intelligence and human marketing: a framework for enhanced customer insights and personalization. Int. J. Multidisc. Res. **6**(6) (2024)

Kumar, V., Gupta, P.: Rural marketing: challenges and opportunities. Paradigm Manag. Res. J. **25**(1), 87–89 (2021)

Malhotra, N.K., Agarwal, J., Peterson, M.: Methodological issues in cross-cultural marketing research: a state-of-the-art review. Int. Mark. Rev. **13**(5), 7–43 (1996)

Markham, A., Buchanan, E.: Ethical considerations in digital research contexts. Encyclopedia Soc. Behav. Sci. 606–613 (2015)

Olson, C., Levy, J.: Transforming marketing with artificial intelligence. Appl. Mark. Anal. **3**(4) (2017)

Rossides, N.: The future of work: managing in the age of AI. Stoic Owl Press (2024)

Salem, F., Mourtada, R.: Social media in the Arab world: Influencing societal and cultural change. Arab Soc. Media Rep. **2**(1), 1–28 (2012)

Teece, D.J., Pisano, G., Shuen, A.: Dynamic capabilities and strategic management. Strateg. Manag. J. **18**(7), 509–533 (1997)

Thakur, J., Kushwaha, B.P.: Artificial intelligence in marketing research and future research directions: science mapping and research clustering using bibliometric analysis. Glob. Bus. Organ. Excell. **43**(3), 139–155 (2024)

Uyanık, M.: Resource-based view in marketing literature. J. Bus. Manag. Stud. **5**(4), 29–39 (2023)

You, N.N., Lou, Y., Zhang, W., Chen, D., Zeng, L.: Leveraging resources and dynamic capabilities for organizational resilience amid COVID-19. South African J. Bus. Manag. **54**(1), 3802 (2023)

Zhang, D., Pee, L.G., Cui, L.: Artificial intelligence in e-commerce fulfillment: a case study of resource orchestration at Alibaba's smart warehouse. Int. J. Inf. Manage. **57**, 102304 (2021)

Digitalization of Consumer and Buyer-Seller Interactions

The Evolution of Buyer-Seller Relationships in the Digital Age: A Philosophical Approach

Erasmia Leonidou(✉)

Cyprus University of Technology, 3036 Limassol, Cyprus
erasmia.leonidou@cut.ac.cy

Abstract. Rapid technological advancements have led to a significant diversification in the media through which buyers and sellers interact. This shift encompasses a broad spectrum of communication channels, extending beyond in-person communication to include technology-driven and multiformat interfaces. Digialization and staggering amounts of information have played a catalytic role in reshaping and reassessing the relationship between informed buyers and sellers. This chapter draws on the literature of Service-Dominant Logic and Service Ecosystems by incorporating a philosophical depth to address the emerging challenges. In the same vein, this chapter tries to shed light on the interplay between individual seller capabilities and broader institutional arrangements, emphasizing the need for sellers to act as drivers of both institutional and societal well-being. Finally, this chapter offers further conceptual insights and outlines key challenges while framing the path ahead for continued exploration in the field of sales.

Keywords: Selling · Value Creation · Eudaimonia

1 Introduction

The era we live in constitutes a turning point in the history of mankind. Even though there is a long history on sales research, the technological upgrade and the revolution of digitization call organizations to rethink the selling process. In order to preserve human connection and the fundamental importance of relationships, scholars need to reconceptualize the basic principles of communication and exchange, so as to marry the new with the classic. Such concerns can contribute to the discussion of exploring ways to uphold the universal spiritual and mental needs of humanity (Brodie et al., 2011; Hartmann, Wieland and Vargo, 2018).

For centuries, human societies have been structured around the principles of exchange. Philosophers and academics have explored the nature of exchange as a fundamental human activity. Nowadays, however, technological developments have transformed the way people work and collaborate with significant societal impact (Benk et al., 2024). This chapter is guided by two main questions:

- How does digitalization affect the foundational human behaviors that govern the selling process?

A. K. Zarkada (Ed.): GMA-GAMMA 2025, SPBE, pp. 77–87, 2025.
https://doi.org/10.1007/978-3-031-96930-0_5

- What are the philosophical implications of recent transformations on value creation?

Innovation and technology are reshaping not only the mechanics of sales but also the ethical, emotional, and cognitive dimensions of buyer-seller relationships. These relationships are increasingly mediated by algorithms, big data, and multiformat communication (Bitner, Brown and Meuter, 2000; Fernandes and Oliveira, 2021). The chapter will reassess the philosophical foundations of this evolving interaction and explore the competencies required for salespeople to navigate this ever-changing environment. This chapter focuses on conceptual aspects and the foundations of the phenomenon itself in order to initiate a fruitful dialogue on how marketing scholars can explore related consequences of technological advancements on seller and consumer interactions.

2 Exchange: A Foundational Perspective

Trade and exchange intrigue humanity for centuries. As early as 350 BC, Aristotle has been dealing with these concepts in the context of fulfilling human needs. For Aristotle, fairness in exchange existed when there was mutual benefit and proportionality, contributing to both individual well-being and social harmony (Crisp, 2014). Also, Aristotle in his work "The Art of Rhetoric", he wrote about "persuasion", that encompass the rhetorical strategies of logos, pathos and ethos, which overlap with a number of influence tactics used today by the management and marketing field (McFarland & Dixon, 2019).

In the 18th century, Adam Smith, in a different direction, emphasized the importance of the functioning of the market through individual action, i.e. "It is not from the benevolence of the butcher, the brewer, or the baker that we expect our dinner, but from their regard to their own interest" (Smith, 2002, p. 30) He argued that economic transactions do not require explicit moral criteria of justice, as Aristotle implied, but instead, that transactions can lead to collective welfare through individual action. In other words, the market works best when people pursue their self-interest, as this develops a form of equilibrium in the market, also known as the "invisible hand".

These approaches set the stage for Clark and Mills (1979) who argued that people exist simultaneously in two different worlds. First, people live in a world that social norms prevail, where there is no immediate need to repay someone, and thus reciprocity is not immediately necessary. The relationships created in this world are also labeled as "communal" relationships, and people assume that each individual is genuinely concerned about the welfare of the others. On the other hand, there is a second world dominated by market norms that work very differently. In this world, people form their relationships based on self-reliance, resourcefulness and individualism. The relationships are called "exchange relationships" and members expect mutual benefits and prompt reciprocity. Interestingly, today some companies, take this theory a step further, trying to benefit from the flexibility, concern and willingness that accompanies "communal" relationships. In other words, they attempt to enter the world of social norms, engaging both employees and customers. This is a very difficult task because when social norms collide with a market norm, the social norm disappears, or as Ariely (2009) concluded «for market norms to emerge, it is sufficient to mention money" (p. 74).

These foundational ideas about exchange have evolved throughout the years in response to the societal, technological and market developments. Today, these ideas have integrated with a systemic approach that emphasizes the value creation. More precisely, selling has been recently defined as the interaction that aims to build long term relationships and maintain exchanges that are easy and efficient, while building trust and optimizing gradually the relationship (Hartmann et al., 2018). Academics zoom out from the linear dyadic view to gain a more holistic and realistic perspective that reveals additional external actors and intrafirm stakeholders that facilitate value co-creation (Bolander, 2015).

In conclusion, it seems that humanity transitioned from the philosophical analysis of Aristotle, to the economic thought of Smith and finally to the modern theories of transactions and the differentiation of social and economic relations. While the nature of exchange has evolved, the natural human tendencies behind trade, like the desire, need, trust, and value, still remain consistent (Ganesan, 1994; Morgan and Hunt, 1994; Voorhees et al., 2017). Similarly, the business world has evolved over the years, trying to find ways to create meaningful, mutually beneficial exchanges with their customers. In this concept, marketing conveys value that aligns with customers' ethical and cultural principles, rather than merely promoting a specific service or product (Vargo and Lusch, 2004, 2016). Companies struggle to fulfill their goals while focusing simultaneously on trust, as without trust there cannot be joint commitments, long-term relationships and partnerships (Voorhees et al., 2017).

3 The Transformative Role of Technology in Buyer-Seller Relationships

Technological disruptions and the information revolution have brought about a novel reality in the way buyers and sellers communicate and collaborate with each other (Singh et al., 2019). Today, their communication has shifted largely from in-person communication to technologically enabled interaction (Grewal et al., 2020; Ahearne et al., 2022). As shown in Fig. 1, digital technology has become a vital part of day-to-day activities for many European consumers. This particular research shows that the extensive use of technological devices, not only shows the widespread acceptance of technology by almost all age groups, but also highlights the need for personalized approaches to sales and communication. Multiformat communication refers to the simultaneous use of multiple communication channels to engage with buyers (Moffett, Folse and Palmatier, 2021), which enables sellers to reach a broader audience and connect with buyers on their preferred platforms. Research has shown that personalized multiformat communication can lead to enhanced customer satisfaction, boosted sales, and strengthened customer loyalty (Voorhees et al., 2017).

Nowadays, both, buyers and sellers, have access to a range of digital platforms, including social media and mobile applications, and use various electronic means of communication, including email, websites, online chats, video conferencing and app-based messaging (Khan and Ebner, 2019). These contemporary communication methods offer new ways for information sharing, negotiation, and transactional activities to take place (Mofett et al., 2021). Furthermore, customers are progressively transforming into informed buyers due to online research, reviews, and recommendations. Instead of relying solely on traditional communication methods, such as phone calls and in-person

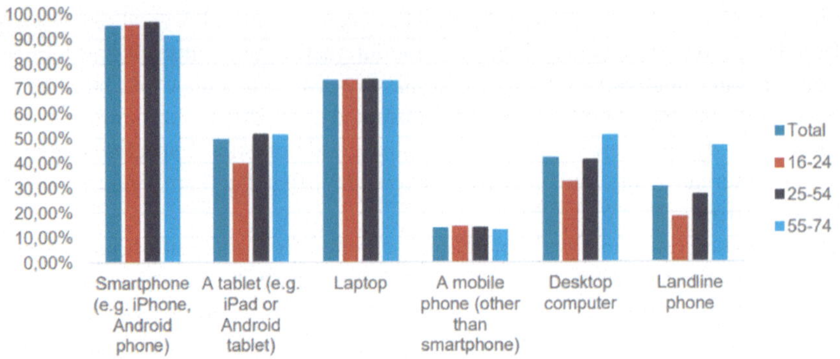

Fig. 1. Ownership and Usage of Communication Devices Among EU Age Groups

meetings, sellers must now adopt a multiformat approach to effectively engage with informed buyers (ibid.).

The growing importance of digital technology for Europeans in the process of buying, selling products and services and customer service is also highlighted in the European Commission's Special Eurobarometer 532 (2023). In particular, the European consumers appear to be increasingly familiar with digital platforms, with a significant 74% of respondents across the EU believing that digital technologies will be vital in their lives by 2030 for engaging in Online Shopping, Selling and Customer Service. This highlights the widespread expectation that digital technology will remain an integral part of consumer activities, which places a strong emphasis on building trust and maintaining customer loyalty.

In the same vein, AI, as a field that combines multiple disciplines to engineer machines that can learn and adapt, make decisions and operate intelligently within the environment (Ferrario, Loi and Viganò, 2020), creates higher expectations for a more personalized and frictionless customer experience (Marinchak, Forrest and Hoanca, 2018). These systems enhance efficiency by adapting autonomously and handling complex tasks. Both employees and customers simultaneously use AI to guide purchasing and other decisions. Inevitably, human agency in the buyer-seller relationship has shifted (Hartmann et al., 2018). Traditionally, sellers used to leverage a number of fundamental human tendencies to influence consumer's purchasing decision. For example, responding to consumers' impulse buying, sellers would offer immediate incentives. Or in another context, sellers would simplify choices or highlight key features of a product to lead buyers towards a more predictable behavior. Today, in-person communication is reduced but more sophisticated, and still needs to respond to the fundamental human tendencies that still remain the same (Marinchak, 2018).

4 Selling – Contemporary Theoretical Grounding

For decades, marketing scholars have tried to uncover the determinants of sales performance. A couple of decades ago, Anderson (1996), had identified a number of forces that impact selling, like the behavioral forces, i.e. the evolving customer expectations; the technological forces, i.e. the various electronic sales channels; and the managerial forces. In the same vein, contemporary sales literature has identified a number of realities that influence the modern sales process, like for example the technological advancements, customer access to information and their increasing demands (Hartmann et al., 2018). Nowadays, these employees often manage projects, develop strategic partnerships and even act as general managers (ibid.). In the same vein, Rapp et al. (2017) formalized the concept of "sales-service interface", to highlight that salespeople take on various roles and are under tremendous pressure while engaging with both sales and service activities. They regularly play an ambidextrous role, responding to all the aforementioned tasks, regardless of their formal position (ibid.).

Based on the theory of Service-Dominant Logic, buyer-seller relationship is collaborative in nature (Vargo and Lusch, 2004). Buyers actively contribute to this relationship by providing feedback, demanding a more personalized service, and simultaneously engaging and interacting through digital platforms. Consequently, selling is not considered a simple exchange based on transactions and transferring tangible goods. On the contrary, exchange is about delivering value through service, which becomes central to the notion of exchange (ibid.), and thus, service becomes the vehicle for value co-creation.

Another important concept introduced in literature is the "service ecosystem", which implies that selling takes place in a contextual infrastructure that involves a complex and dynamic network of actors (Vargo and Lusch, 2016). In other words, sellers' function within shared institutional arrangements, like rules, meanings, symbols and practices that affect the value co-creation (ibid). These shared institutional arrangements govern the relationships and ensure trust and stability. Sellers need to synthesize various information and coordinate interactions with multiple actors and various resources (ibid.). All these evolutions generate the contemporary knowledge, skills, competencies and abilities (KSCAs), sellers need in order to foster cooperation, optimize relational exchanges, and create and deliver value (Rapp et al., 2017).

Moreover, recent research shed light on the importance of in-person sales interaction within the context of omnichannel selling (McFarland and Dixon, 2019; Kim and McFarland, 2024). In-person interactions facilitate the identification of customer needs and adaptation of sales practices and behaviors accordingly, known as "adaptive selling". However, despite the extended research on this topic, there are still important gaps in the literature of adaptive selling (Hochstein et al., 2019; Kim and McFarland, 2024).

5 Theoretical Enrichment for Future Research

All these synthesize a complex and dynamic environment in which sellers cannot rely on the technological advancements to fully replace their ability to foster trust, respond to subtle cues and strengthen ongoing relationships (Hochstein, 2019). The changes that constantly come from the macro-environment, like the various legislations and regulatory frameworks, along with the dynamics of intra-organizational network, inevitably affect sellers' performance (Bolander et al., 2015). Accordingly, the seller must be a person who is aware and willing to be constantly educated about the developments that affect the nature of his or her work (Rapp et al., 2017).

This reality calls for a more in-depth approach to the nature of selling (ibid.). It is Aristotle's contention that every human activity aims at the creation of some sort of "good", which in turn serves some other "good", and so on. This chain of "goods", however, should have "an ultimate end". Aristotle considers "eudaimonia" to be the activity of living ethically, guided by moral virtues and practical wisdom, and achieving the highest level of intellectual activity (Broadie, 1993). In selling we consider the ultimate end to be the value creation (Voorhees et al., 2017). Aristotle urges as a first and necessary step the determination of each person's "ergon", or in other words function (Nagel, 1972; Crisp, 2014), so as to be able to measure how well they perform. Continuing with our analogy, in the case of sellers, the reference point is the clarification of their role, tasks and responsibilities, which will showcase at a later stage, the quality level of their performance. "Eudaimonia" is a concept that must be cultivated in sellers, in the name of which they must ultimately regulate their behavior. Selling becomes a contribution to the broader service ecosystem and sellers ensure that every single interaction enhances the institutional well-being and value (Voorhees et al., 2017).

Sellers are expected to be agile and adopt to this transforming landscape (Rapp et al., 2017). Inevitably, sellers are expected to acquire a set of evolving competencies that will help them succeed in the digital world (ibid.). For example, sellers are expected to understand and handle vast amount of data and critically engage with it. This competence of critically approaching the data is a contemporary form of "phronesis", or, in other words "practical wisdom", which according to Aristotle, is 'a true and reasoned state of capacity to act with regard to the things that are good or bad for man" (Noel, 1999, p. 273). Today this competence is about the ability to apply reason to experience to make an ethically sound decision. Modern sellers are expected to understand the data and critically interpret them in order to foster trust and cultivate longterm relationships (Rapp et al., 2017).

In an attempt to provide a conceptual bridge between philosophy and business practices, we introduce the concept of "Eudaimonic Selling", which emphasizes the ethical practices and the mutual value creation in the pursue of long-lasting relational exchanges. In this context, selling resonates with modern sustainability and contemporary regulations emphasizing values, honesty, fairness and empathy, that promote institutional flourishing and human well-being. A formal requirement of "eudaimonia" is "self-sufficiency", which means that "eudaimonia" does not depend on only one element but rather on the integration of various factors that relate to morality, wisdom and action (Heinaman, 1988). In the context of selling, "self-sufficiency" translates to the holistic approach towards meeting the customer's needs and ensuring that the overall experience will also meet their evolving expectations.

As illustrated in Fig. 2, "Eudaimonic Selling" represents a dynamic interplay between the foundational pillars of "eudaimonia" and the contemporary theory of selling. On the one hand, there is "virtue", which is about the ethical principles (Nagel, 1972), and "self-sufficiency" that refers to the sustainability required for long-term relationships (Heinaman, 1988). On the other hand, there is "institutional arrangements", identified by contemporary academics, that enable value co-creation (Vargo and Lusch, 2016), which is exactly the intersection of the circles. This conceptual model provides a holistic

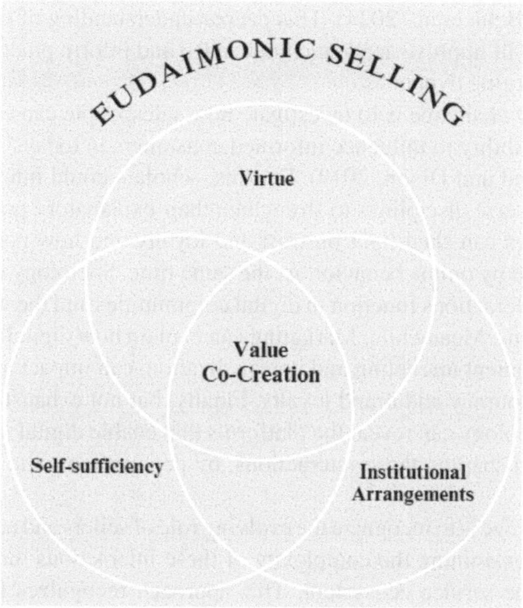

Fig. 2. Conceptual Framework of «Eudaimonic Selling»

approach towards the complex dynamics of contemporary selling practices, highlighting the importance of ethical judgment and agility in the evolving environment.

6 Framing the Future: Challenges and the Road Ahead

As noted at the outset of this chapter, the aim is to explore the evolution of Buyer-Seller Relationships in the Digital Age and to encourage a productive dialogue on how technological mediation has transformed these interactions and how human behavior adapts to the transforming landscape.

First, in an effort to understand buyer-seller interaction within the new paradigm, we need to adopt a reflective thoughtful approach in interpreting the technological advancements. Despite the fact that it is hard to predict upcoming technological breakthroughs, it is likely that more advanced AI systems could bring about impactful social change, since the core purpose of the field of AI has long been to create a machine capable of exhibiting humanlike "common sense" (Gruetzemacher and Whittlestone, 2022). Thus, we need to discover the links between different elements in buyer-seller interactions and to uncover ways of managing these relationships in the digital environment (Kim and McFarland, 2024). For example, up until today we have known "Trust" to be the reliance in another person's voluntarily accepted responsibility to recognize and safeguard the interests of all parties engaged in a joint endeavor or financial exchange (Hosmer, 1995). Today, this definition still applies to interpersonal and organizational context, as "Trust" is still the cornerstone of effective communication. On the other hand, when it comes to the interaction with technology, trust is widely regarded as instrumental for efficiency and effectiveness (Benk et al., 2024). This deeper understanding of the theories, methods and contexts, will improve academic knowledge and inform practitioners on how to adopt to these emerging dynamics.

The second key challenge is to investigate how salespeople can establish a basis of reliability and credibility to influence informed consumers in todays' challenging environment (McFarland and Dixon, 2019). For this, scholars could integrate theories and constructs from diverse disciplines to strengthen their explanatory power. For instance, Behavioral Sciences can shed light on trust and loyalty and how perceptions of these concepts are shaped by online behavior. At the same time, Sociology can help us understand how social interactions function in digital communities and the way these relationships affect decisions. Meanwhile, Marketing can explain how digital marketing tactics, like for example content marketing and personalization, can impact loyalty, and further analyze customer journey and brand loyalty. Finally, but not exhaustively, Information Systems and Technology can reveal the platforms that enable digital interaction, as well as the role of AI in shaping these interactions, by personalizing the whole experience and building trust.

The third challenge is to recognize the evolving role of sellers and address the changes by systematically examining the complexity of these interactions through a multilevel approach within the service ecosystem. This approach recognizes the importance of organizational dynamics in supporting sellers for successfully adapting to the new environment. It also offers grounded theoretical explanations that extent our knowledge on the evolving role of sellers, including the scope of knowledge, skills, behaviors and abilities that are essential to perform these expanding tasks (Bolander, 2015).

The fourth challenge is to place sellers at the forefront of Environmental, Social and Governance (ESG) compliance. Sellers, as the boundary spanners who implement the companies' strategies, could drive such initiatives forward by acting as ambassadors of a company's ESG values and goals. This approach is both timely and relevant since the European Union has recently intensified its focus on ESG practices through a number of regulations and guidelines, like for example the Corporate Sustainability Reporting Directive, under which companies need to disclose their ESG practices (Directive EU 2022/2464). These regulatory requirements impact the selling process as sellers are now expected to align their offerings with ESG practices and support sustainable purchasing decisions. Today, this is more likely than ever, as digital marketing communication and tools effectively convey information about the organizations' values (Jahdi and Acikdilli, 2009), like sustainability, thus, can play a key role in helping sellers promote environmental protection. In other words, research should explore the possibilities of investing in the digital connectivity, to facilitate the pursuit of sustainable approaches, and simultaneously invest in equipping employees with the knowledge and tools to integrate ESG values into their practices (Directive EU 2022/2464). For example, sellers can guide consumers by raising awareness about the organizations' socially responsible products.

In conclusion, research should examine the methods and processes used to acquire the required KSCAs, needed to accelerate sellers' performance in the increasingly interconnected world. Responding to these fundamental realities, companies need to focus on the entirely new set of KSCAs and implement a more sophisticated method in the recruitment of sellers, as well as the training to support these dynamic capabilities. Importantly, organizations need to acknowledge and embrace the complexity that accompanies this role, by integrating sophisticated performance management tools to manage and maintain employees' motivation (Rapp et al., 2017). It is in this context that the concept of "Eudaimonic Selling" is proposed, which, although it leads to an effortful growth, ultimately leads to a sense of purpose and personal expressiveness (Ryan and Deci, 2001).

7 Conclusion

In this chapter, the focus lies on the complexities of buyer-seller interactions, considering important concepts, such as human communication and trust. While simplifications are often tempting, they do not capture the depth and richness of these relationships that are currently mediated by technology. This philosophical, multidisciplinary and multi-level approach has several empirical implications for future studies on this area. Achieving a deep level of understanding will enable a more strategic approach towards successful interactions in the digital environment, by helping salespeople to navigating the complexities while maintaining the fundamental humanistic principles. Lastly, scholars need to shed light on the changing role of sellers by investigating contemporary communication strategies and tactics to effectively engage with informed buyers and cultivate strong relationships.

References

Adams, J.S.: Towards an understanding of inequity. J. Abnorm. Soc. Psychol. **67**(5), 422–436 (1963)

Ahearne, M., Atefi, Y., Lam, S.K., Pourmasoudi, M.: The future of buyer– seller interactions: a conceptual framework and research agenda. J. Acad. Mark. Sci. **50**, 22–45 (2022)

Anderson, R.E.: Personal selling and sales management in the new millennium. J. Pers. Sell. Sales Manage. **16**(4), 17–32 (1996)

Ariely, D.: Predictably Irrational: The Hidden Forces that Shape Our Decisions. HarperCollins Publishers (2009)

Benk, M., Kerstan, S., von Wangenheim, F., Ferrario, A.: Twenty-four years of empirical research on trust in AI: a bibliometric review of trends, over-looked issues, and future directions. AI Soc. 1–24 (2024)

Bitner, M.J., Brown, S.W., Meuter, M.L.: Technology infusion in service encounters. J. Acad. Mark. Sci. **28**(1), 138–149 (2000)

Body of European Regulators for Electronic Communications (BEREC). Analysing EU consumer perceptions and behaviour on digital platforms for communication: Analysis report (BoR (21) 89) (2021). https://www.berec.europa.eu/en/document-categories/berec/reports/analysing-eu-consumer-perceptions-and-behaviour-on-digital-platforms-for-communication-analysis-report

Bodó, B.: Mediated trust: a theoretical framework to address the trustworthiness of technological trust mediators. New Media Soc. **23**(9), 2668–2690 (2021)

Bolander, W., Satornino, C.B., Hughes, D.E., Ferris, G.R.: Social networks within sales organizations: their development and importance for sales-person performance. J. Mark. **79**(6), 1–16 (2015)

Broadie, S.: Ethics with Aristotle. Oxford University Press, Oxford (1993)

Brodie, R.J., Hollebeek, L.D., Jurić, B., Ilić, A.: Customer engagement: conceptual domain, fundamental propositions, and implications for research. J. Serv. Res. **14**(3), 252–271 (2011)
Clark, M.S., Mills, J.: Interpersonal attraction in exchange and communal relationships. J. Pers. Soc. Psychol. **37**(1), 12–24 (1979)
Crisp, R. (ed.): Aristotle: Nicomachean ethics. Cambridge University Press, Cambridge (2014)
European Commission. Special Eurobarometer 532: The Digital Decade. Directorate-General for Communication (2023). https://europa.eu/euroba-rometer/surveys/detail/2959
European Parliament and Council. Directive (EU) 2022/2464 of the European Parliament and of the Council of 14 December 2022 amending Regulation (EU) No 537/2014, Directive 2004/109/EC, Directive 2006/43/EC and Directive 2013/34/EU, as regards corporate sustainability reporting (2022). https://eur-lex.europa.eu/legal-content/EN/TXT/HTML/?uri=CELEX:32022L2464
Fernandes, T., Oliveira, E.: Understanding consumers' acceptance of automated technologies in service encounters: drivers of digital voice assistant's adoption. J. Bus. Res. **122**, 180–191 (2021)
Ferrario, A., Loi, M., Viganò, E.: In AI we trust incrementally: a multi-layer model of trust to analyze human-artificial intelligence interactions. Philos. Technol. **33**, 523–539 (2020)
Ganesan, S.: Determinants of long-term orientation in buyer-seller relationships. J. Mark. **58**(2), 1–19 (1994)
Grewal, D., Hulland, J., Kopalle, P.K., Karahanna, E.: The future of technology and marketing: a multidisciplinary perspective. J. Acad. Mark. Sci. **48**, 1–8 (2020)
Gruetzemacher, R., Whittlestone, J.: The transformative potential of artificial intelligence. Futures **135**, 102884 (2022)
Hartmann, N.N., Wieland, H., Vargo, S.L.: Converging on a new theoretical foundation for selling. J. Mark. **82**(2), 1–18 (2018)
Heinaman, R.: Eudaimonia and self-sufficiency in the "nicomachean ethics." Phronesis **33**(1), 31–53 (1988)
Hochstein, B., Bolander, W., Goldsmith, R., Plouffe, C.R.: Adapting influence approaches to informed consumers in high-involvement purchases: are salespeople really doomed? J. Acad. Mark. Sci. **47**, 118–137 (2019)
Hosmer, L.T.: Trust: the connecting link between organizational theory and philosophical ethics. Acad. Manag. Rev. **20**(2), 379–403 (1995)
Jahdi, K.S., Acikdilli, G.: Marketing communications and corporate social responsibility (CSR): marriage of convenience or shotgun wedding? J. Bus. Ethics **88**, 103–113 (2009)
Khan, M.A., Ebner, N.: (eds.): The Palgrave Handbook of Cross-Cultural Business Negotiation. Palgrave Macmilla (2019)
Kim, Y., McFarland, R.G.: Are you looking for something specific or just looking around? Adaptive selling on the basis of customer shopping goals in retail sales. J. Acad. Mark. Sci. **52**, 1780–1804 (2024)
Marinchak, C.M., Forrest, E., Hoanca, B.: Artificial intelligence: redefining marketing management and the customer experience. Int. J. E-Entrepreneursh. Innov. (IJEEI) **8**(2), 14–24 (2018)
McFarland, R.G., Dixon, A.L.: An updated taxonomy of salesperson influence tactics. J. Pers. Sell. Sales Manage. **39**(3), 238–253 (2019)
Moffett, J.W., Folse, J.A.G., Palmatier, R.W.: A theory of multiformat communication: mechanisms, dynamics, and strategies. J. Acad. Mark. Sci. **49**, 441–461 (2021)
Morgan, R.M., Hunt, S.D.: The commitment-trust theory of relationship marketing. J. Mark. **58**(3), 20–38 (1994)
Nagel, T.: Aristotle on eudaimonia. Phronesis **17**(3), 252–259 (1972)
Noel, J.: On the varieties of phronesis. Educ. Philos. Theoryeory **31**(3), 273–289 (1999)

Rapp, A.A., Bachrach, D.G., Flaherty, K.E., Hughes, D.E., Sharma, A., Voorhees, C.M.: The role of the sales-service interface and ambidexterity in the evolving organization: a multilevel research agenda. J. Serv. Res. **20**(1), 59–75 (2017)

Ryan, R.M., Deci, E.L.: On happiness and human potentials: a review of research on hedonic and eudaimonic well-being. Annu. Rev. Psychol.l. **52**(1), 141–166 (2001)

Singh, J., et al.: Sales profession and professionals in the age of digitization and artificial intelligence technologies: concepts, priorities, and questions. J. Pers. Sell. Sales Manage. **39**(1), 2–22 (2019)

Smith, A.: An inquiry into the nature and causes of the wealth of nations. Readings in Econ. Sociol. 6–17 (2002)

Taddeo, M., Floridi, L.: The case for e-trust. Ethics Inf. Technol. **13**(1), 1–3 (2011)

Vargo, S.L., Lusch, R.F.: Evolving to a new dominant logic for marketing. J. Mark. **68**(1), 1–17 (2004)

Vargo, S.L., Lusch, R.F.: Institutions and axioms: an extension and update of service-dominant logic. J. Acad. Mark. Sci.i. **44**(5), 5–23 (2016)

Voorhees, C.M., et al.: Service encounters, experiences and the customer journey: defining the field and a call to expand our lens. J. Bus. Res. **79**, 269–280 (2017)

The Social Shift: How Social Media Revolutionized Consumer Behaviour Research

Yioula Melanthiou(✉)

Communication and Marketing, Cyprus University of Technology, 30, Archbishop Kyprianou Str., 3036 Limassol, Cyprus
yioula.melanthiou@cut.ac.cy

1 Introduction

Let's think of a world where every thumbs-up, comment, or share carries the story of consumer desires and changes how companies interact with their customers. This is just how the contemporary advertising world looks like today because of social media's miraculous power. Gathering information directly from people through face-to-face conversations or manually filling out questionnaires seems like a thing of the past. Nowadays, a vast amount of real-time data that is unrivalled for its scope is open to today's marketers; this helps them to understand the purchasing behaviour of their customers better and more objectively. Hence businesses now approach gaining insights differently enabling them to develop new data-driven innovative marketing techniques.

Social media has changed the process of data collection and at the same time altered the dynamics between brands and consumers. Facebook, Instagram or TikTok are platforms that modern marketers cannot do without. On these sites, they can study audience behaviour, engage directly with them and even influence their choices - something that was not possible twenty years ago. It represents a digital transformation where interaction and consumer influence reign and supreme information sharing forms core marketing strategies that are crafted and implemented differently in this new environment.

The chapter explores critically the development of consumer behaviour research, contrasting traditional methods with social media-driven approaches, and evaluating the implications for marketing strategy, data quality, sustainability, and ethical behaviour in the era of artificial intelligence. It further examines the move from what may currently be considered as 'old-school' consumer behaviour research to one which now encompasses social media tools. Social media plays a key role in driving marketing innovation and will continue to do so. Companies however will need to understand both its potential and strength but also the ethical challenges associated with employing digital consumer data effectively and lawfully.

2 An Overview of Traditional Consumer Behaviour Research

Understanding the behaviour of consumers has always been a marketing pillar, varying with changes in technology and society. The early attempts at research date back to the late 19th and early 20th centuries and were based on simple observations and basic

A. K. Zarkada (Ed.): GMA-GAMMA 2025, SPBE, pp. 88–103, 2025.
https://doi.org/10.1007/978-3-031-96930-0_6

methods. The main goal of such studies was to determine what people bought, why they did so, and how they made their choices when making purchase decisions. Over time, this has developed into a more systematic field embracing both qualitative and quantitative methodologies.

By the end of the 20th century, it became structured consumer behaviour research. Surveys, focus groups and in-store observation were starting to become more common. By using such methods, marketers could interact directly with clients in order to have credible information on their preferences and motivations. Surveys were employed for structured measurable data while focus groups allowed for deeper exploration of consumer attitudes and emotions. On the other hand, observations in retail outlets gave immediate information on purchase behaviours as well as decision- making processes.

These methodologies played a critical role in setting up a basis for contemporary marketing strategies. They enabled enterprises to obtain clearer insights into customer needs which were used to develop products as well as campaigns that resonated with targeted audiences. For instance, during the 1950s household brands such as Campbell's used surveys to measure consumer satisfaction and preferences thus improving their products while simultaneously enhancing brand loyalty.

Despite being useful traditional methods of conducting consumer behaviour research, these studies had some inherent drawbacks. Analysis could only take place based often on small samples sizes because most studies focused on particular demographic groups or geographical areas. Consequently, this limited the generalizability of the findings hence questioning their relevance and reliability to other populations. Traditional research methods were also time consuming and expensive to implement. For instance, it would take a lot of time and money to conduct focus groups or distribute surveys and analyse the data collected from them. Information that was obtained through this process would become outdated by the time it was acted on since there was often a significant lag between data collection and actionable insights. For example, if a study was conducted for several months, results from such study might no longer be applicable due to changes in the market environment or consumer preferences that are always evolving. Another major limitation of this type of research was the over reliance on self-reported information. Individual customers unintentionally forget some details from their past experiences or dishonestly answer questions asked with an aim of presenting themselves in good light. For instance, in one survey an individual might claim that he prefers eating a lot of healthy food but in reality, buys snacks full of sugar and fats which clearly indicates inadequacy between what one says and does. Due to this discrepancy between what people say about themselves versus what they actually do, traditional research lacks validity.

In the 20th century, prominent corporations like Coca-Cola heavily depended on traditional means of research for guiding decisions. For example, taste tests within stores were used by Coca-Cola to get insights at a local level and subsequently define its product category. Such tests however, did not consider broader cultural or regional issues, resulting in marketing strategies that could not be universally accepted. For example, despite positive feedback from focus groups carried out, the introduction of New Coke in the 1980s witnessed a massive resistance from loyal customers who accused the company of losing touch with their tastes, forcing them to go back to the original taste of Coca-Cola only months after. This case is often referred to as 'the biggest marketing blunder of all time'.

Likewise, Procter & Gamble (P&G) utilized focus groups to comprehend consumer requirements and preferences concerning their products like Tide washing powder and Pampers diapers. These studies were confined to particular regions which made it difficult for them to anticipate general market patterns or future consumer needs. Recognising the limitations of their traditional consumer research methods P&G moved to using online surveys and social media sentiment analysis to track consumer opinions in real time, enabling faster responses to emerging trends. In this rapidly shifting environment, these examples serve as clear illustrations of what traditional methods suffer from in terms of inefficiency in considering developing markets.

Traditional research methods certainly do have some limitations; but, they are necessary especially in the deeper qualitative inquiries that one may wish to conduct. For example, focus groups or ethnographic studies can reveal the emotional drives behind choice behaviour, sensitivities to cultural transmission, or the narratives of consumer loyalty that are not otherwise parsed from quantitative data. These traditional methods have however demonstrated how slow, small-scale and exclusive they are leading marketers to realize the urgent need for faster, scalable and more participatory models. The digital age has given a breakthrough to researchers who had always wanted to overcome drawbacks of traditional methods.

3 Social Media in Consumer Behaviour Research

The beginning of the 21st century witnessed the genesis of Facebook, Twitter (now X) and LinkedIn as new avenues for consumer behaviour study through social media. By capturing real-time data on an unprecedented level, these sites revolutionized how researchers could view individuals' opinions, choices and discussions at any given moment across the world. Not only did this bridge all existing communication gaps that had been left by more conventional methods it also brought in some new realms for understanding consumers including peer influence, emotion analysis as well as tracking their actions online.

For consumer behaviour research this marked the beginning of a new era so great in terms of speed of data acquisition and data accuracy owing to digitization. Businesses could keep up with what was trending and experiment their publicity materials while changing their course if needed when they missed the point in real time. Nike, for example, and Starbucks use platforms like X and Instagram to uncover emerging trends before launching focused ads that capture particular sections of their market.

3.1 Social Media as a Strategic Marketing Tool

Social media has become one of the strongest platforms from which businesses communicate and track audience behaviours against market changes. The evolution of consumer behaviour research has mirrored the wider trajectory of marketing practices, shifting from localized, resource-intensive to dynamic data-driven strategies. However, traditional methods have set the stage for understanding consumer preferences, which necessitate innovation as evident by their constraints. On the other hand, the emergence of digital technologies and social media completely altered this sphere allowing firms to reach their audience unprecedentedly and follow current market trends at all times. This transition not only addresses the difficulties encountered using customary techniques but also provides a pathway towards future progressions within consumer behaviour research.

Social media data is becoming more vital to marketers as they design ad campaigns and establish customer insights. Such a change from former methods to using online information found in social networks, enables businesses to streamline their marketing strategies and optimize their targeting strategies avoiding wastage of resources enhancing performance and reducing costs which will be discussed further on in the chapter. Some studies have empirically validated how prevalent such a change is towards marketing techniques. The study by Li et al. (2020) asserted that social media is indeed an important marketing tool but many marketers do not have enough information or training on how to translate social media data into actionable insights. Through their work, they note that there is a need for organizations to gain the necessary skills which will enable them to utilize online tools effectively and to its maximum capability. This major transition prepared the grounds for innovations of data gathering and usage, most prominently the rise of real-time research methodologies.

3.2 A Paradigm Shift in Real-Time Data Acquisition

There have been several changes in how firms collect and utilise information from social media, enabling them to gather real-time data facilitating agile marketing tactics. Social media platforms offer corporations immediate access to views, inclinations and conducts of the consumers. Through this, they are able to react fast towards market trends and client demands. Brands use social media to communicate directly with their customers creating a feedback loop that is not seen in traditional market research means. This type of conversation results in more accurate and up-to-date knowledge. For researchers, it is possible to see how consumers are interacting amongst themselves as well as what they like or what they find attractive when making purchases. Social media enables marketers to access the voices of customers, thus facilitating a deeper comprehension of their thoughts and feelings which might be very different in sentiments from conventional ways.

Figure 1 shows how companies collect and use real-time data from social media sites. Companies can use X Analytics for example, to amass all trending hashtags and feelings (data capture), AI powered tools can help in distinguishing happy or sad comments about the product (data analysis) and subsequently may opt to tune advertising campaigns more towards aspects that are more relevant to their audience (application).

3.3 How Social Media Influences Customer Participation and Brand Equity

Godey et al. (2016) inquired how social media marketing activities influence brand equity and consumer behaviour. Luxury brands, for instance, are increasingly using social media to engage with consumers and build brand equity as evidenced from their research findings suggesting that this market segment heavily relies on social media data

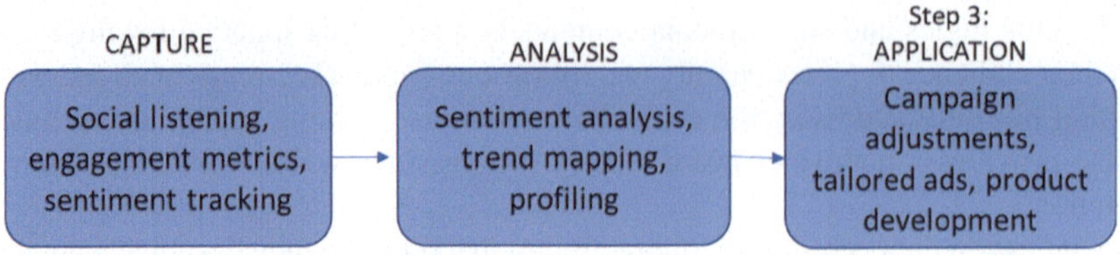

Fig. 1. The Real-Time Data Collection Process

to inform marketing strategies. This gives greater credence to the role played by social media in shaping consumer perception and driving engagement levels.

Consumer behaviour is highly influenced by social media. A better understanding of these behavioural manifestations in social media context is crucial for organizations as consumer interactions within social networks provide insights about customer experiences and engagement indicators. With the unlimited access to any type of information, people appear to trust their friends and people they know better than any other information source. Professionals in marketing will therefore find it essential to employ social networks, and use brand ambassadors or brand influences to assist in their marketing efforts.

4 Optimizing Marketing Costs Through the Use of Social Media Analysis

Utilizing social media for consumer research can be considered less expensive than taking surveys or conducting focus groups as many businesses have started to realize. Social media analytics also provide companies enough information thereby resulting in a reduction of wastefulness involved in targeting. Moreover, they can also rely entirely on strategies related exclusively toward their target audience. Corporations perceive social networks as instruments for gathering customers' ideas and opinions about products or services offered by them (Koesharijadi et al., 2022) and more business are investing resources in hiring professionals to manage their social media not just for promotion purposes but also for data collection. This indicates that marketing campaigns are becoming more data-driven than ever before.

Social media can also be used to accelerate new product acceptance. By gathering precise information from target audience, clear needs for products or services are uncovered, ensuring that new product offerings will directly satisfy these needs. Companies can spend less money and time on gathering traditional marketing research but rather fuel resources towards more strategic initiatives.

Location-based methods of the past have been overtaken by social media platforms where researchers can access numerous global users. Brands can use tools like social listening to understand how people feel about their products all over the world including changing flavours based on global ideas emanating from these platforms. An example is Starbucks where they monitor flavour trends across various social media platforms in order to change their products accordingly.

5 Using Artificial Intelligence and Predictive Analytics in Consumer Research

With historical data combined with real-time information, organizations are able to create highly accurate forecasts using predictive analytics through the merging of artificial intelligence (AI), machine learning (ML), traditional statistics techniques. Organizations like Netflix use predictive analytics to better service their clients. For instance, Netflix uses AI algorithms for analysing the viewers' favourite genres and shows so that it can suggest movies and TV series based on individual users' tastes. Insights gathered through online insights can also give a direction for production in line with the public's interest. Like Netflix, Amazon uses predictive analytics in its supply chain management processes, forecasting sales trends, and recommending certain products to customers depending on their browsing and shopping history. These examples indicate how effectively large-scale operations apply predictive analytics in practical terms.

While bigger firms dominate conversations around forecasting analytics, small players in specialized markets are increasingly adopting such tools as a way of gaining a competitive edge. Unlike large organisations, smaller companies have often far less financial resources making predictive analytics a very crucial tool to maximize return on investment (ROI). Take for example a small fashion house which employs prediction analysis during busy times of the year. Combining historical transaction information with social media data assists the store owner to understand which fashion trends will be popular in future seasons hence stocking appropriately and avoiding excess or shortage. Tools like Google analytics, combined with machine learning algorithms are employed in order to enable these small companies compete with larger retailers.

Pricing strategies for restaurants are becoming more dynamic through predictive analytics. Peak dining times can be predicted by analysing bookings, weather data as well as local events enabling pricing or menu adjustments according to the situation. This results in increased revenues but also better client satisfaction because there is a connection between customer demand patterns and service delivery methods.

Predictive analytics allows access to advanced decision-making tools, permitting both large and small businesses to anticipate consumer needs with greater accuracy, optimize resource allocation, such as inventory management and staffing, and enhance marketing ROI by targeting the right audience with personalized campaigns. By using social media data, advanced data analysis techniques, such as predictive modelling and social network analysis, can improve consumer behaviour understanding enabling marketers to design more effective marketing strategies (Liu, 2024).

6 Advanced Targeting Strategies in Social Media Marketing

Social media platforms offer advanced targeting capabilities that allow companies to personalize their marketing messages for specific customer groups. This approach ensures that companies do not promote their offerings to uninterested audience hence saving on resources (Mason et al., 2021). Companies can capture market trends and customer preferences from the way people interact on social platforms and use the insights to make their campaigns more personal and effective. According to Kumar (2016), businesses can create content that resonates with what their audience want so as to increase the rates of engagement and conversion.

Marketing strategies have evolved due to the presence of social media platforms with sophisticated targeting features that enable organizations to target particular customer groups with great accurateness. Among other things, it allows a firm to personalize marketing messages by analysing demographic facts about their clients like age or geographic information while taking into consideration consumer behaviour on such platforms.

Data collection is the first step in advanced targeting. Facebook, Instagram, LinkedIn and TikTok are examples of platforms that keep vast amounts of data related to the users including age group, location, interest areas or even past buying history information. This information is later segmented through computer algorithms which group similar individuals into separate identifiable categories or classes. For instance, there may be people who visit health and wellness pages most frequently than others therefore they belong to a category known as "health and wellness". Therefore, these users can be targeted by companies offering products and service in this field.

Social media platforms monitor different user actions including likes, shares, and comments among others, to predict their probable behaviour and consequently understand their intent. This is done to ensure ads are displayed for users who are likely to engage with such content thus increasing conversion rates for those campaigns. One such company is Sephora which uses data gathered from behavioural patterns exhibited through its apps. Users are being served personalized promotions about new releases or discounts made available in line with their preferences. If a user has consistently been browsing skincare, then a promotional voucher of a particular skincare product would pop up or an informative advert showing '2024 Best Skincare Products' would be displayed. The aim is not only to prompt purchases but also to remind customers that they are experts in the field of beauty. Similarly, AirBnb leverages on a social media targeting approach by displaying pertinent ads to users experiencing inquiries concerning different places. In case a visitor is researching about Paris tours for example, an illustration of promotional Paris based rental homes accompanied by travel advice to these locations might appear in his/her timeline. By so doing, it has been able to ensure steady bookings and foster customer relationships.

In addition, businesses can use geo-targeting to concentrate their marketing efforts on users from certain locations. Local firms and event organizers may find regional targeting to be beneficial. For example, a coffee shop might run an Instagram campaign specifically targeting the five-mile radius around it to promote a new seasonal drink. A nearby pastry shop could also exploit Facebook's targeting capabilities by sending messages through the platform to neighbours who show interest in cakes or cookies. By advertising about fresh breads, this bakery could draw customers hence increasing sales without having to incur so much on general campaigns. Figure 2 shows the overlap between targeting strategies and corresponding benefits within social media promotion.

6.1 Benefits of Targeted Marketing

Businesses can create ads that feel more personal and engaging enhancing relevance by tailoring content to match consumer preferences. This makes it more likely for consumers to respond positively because personalized marketing fosters a sense of connection and

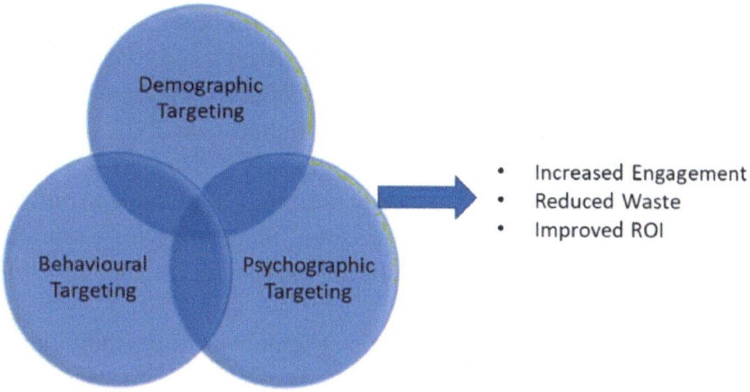

Fig. 2. Targeting Strategies and Benefits

trust. Customers engage with brands through social media platforms. Consequently, Yadav and Rahman (2018) discovered that social media marketing activities have a positive effect on customer loyalty via customer engagement. From this we can deduce that marketers are not only using social media data in their campaigns but are also measuring the effectiveness of these campaigns in terms of customer loyalty and engagement. Furthermore, findings from Kumar (2016) showing that personalized marketing campaigns have significantly higher engagement rates and conversions are important to bear in mind. As a result, targeting audiences with specific ads performs better than generic ones do because they address individual needs and preferences of the audience.

Through the targeted nature of social media advertising, marketers are able to avoid ad wastage risk by focusing on potential clients who can convert thus allocating cost effective resources. As observed by Rahmani (2024), information derived from analysing social media engagement metrics offers valuable insights into customer preferences enabling companies to refine their marketing strategies and avoid wastage. This results-oriented approach ensures that marketing efforts will be efficient and customer-oriented. The efficiency is particularly important for small and medium scale businesses which are budget constrained. When businesses get their target group correctly and use the resources appropriately it becomes possible for them to always attain the highest possible return on investment (ROI). Going forward, it is more likely that the accuracy of targeting will become unmatched by virtue of the emerging technological developments, as discussed in the succeeding sections.

6.2 The Future of Target Marketing

Marketing targeting will not be the same if AI and ML continue to thrive in the world's technological fields. These new technologies enable brands to interact with their customers differently by providing highly personalized marketing campaigns that predict consumer behaviour. More advanced predictive analytics will allow businesses to predict customer needs and preferences with unprecedented accuracy. Predictive models based on vast data sets including website visits, purchasing histories, social network interactions as well as other factors like climate change or economic cycles are able to predict future behaviour patterns of individuals regardless of any other factors.

Moreover, companies will need to stop focusing on being reactive while building marketing campaigns in response to consumer expressed issues; instead they will focus on creating campaigns that satisfy customer needs before they are spoken about. For example, a fitness brand could, with the use of predictive analytics, identify potential customers by garnering online comments regarding their 'New Year's Fitness Resolutions' comments on social media. This strategy is equally applicable to online businesses which suggest products based mostly on seasonal tendencies for example, timing of graduations, Christmas parties, and so on, thereby adding more value to consumers and improving conversion rates in turn.

At the same time, using predictive models at scale will make individualized hyper-personalization possible. Instead of segmenting target groups into large buckets, organisations would rather produce unique marketing communications that would serve each individual customer best possible. So, for instance, a streaming platform can identify that during prime hours a specific user prefers watching romantic comedies, thus can easily speculate on the genre preference and suggest similar content but can optimise type and timing of suggestions and ad placements of that particular user.

As augmented reality (AR) and virtual reality (VR) technologies become more main stream they will play an essential part in targeted marketing offering immersive experiences far surpassing traditional adverts. AR allows for the visualization of items in their natural environment in a way that wasn't possible before when dealing entirely with digital advertising. AR/VR capabilities are already being incorporated into advertising strategies of social media platforms. For example, a furniture store like IKEA, has an AR app that allows its customers to see how a sofa would look like in their living room through augmented reality. This not only enhances the shopping experience but also decreases potential uncertainties and increases shopping confidence. A travel company, for example, uses VR to carry the client into another realm because it is entirely immersive, enabling the company to develop unique ways of engaging the customers. This entails using 3D perspectives of hotels, attractions and local culture within the 360° image of a destination that even allows them to make booking reservations. Snapchat also offers AR filters enabling a try-on of virtual products like sunglasses or makeup. As AR/VR proliferates over different sectors it will allow brands to come up with vastly captivating as well as personalized ad experiences through AR/VR technologies that connect to individuals.

Additionally, the Internet of Things (IoT) and wearable technologies will enable us collect real-time data about what customers do in their everyday lives through information such as online searches, social media engagement or purchase patterns made from devices all around them. Such devices usually include smart home products like Echo Dot from Amazon and Google Nest Hub that monitor everything from music preferences to daily routines of individuals living within their homes. This data could be used for more personalized advertising. For instance, a coffee brand can target users who set their coffee machines to brew at 6:00 a.m. by alluding to special promotions on premium beans or accessories. At the same time fitness wearable devices provide insights into human habits and lifestyles making it possible to conduct goal-oriented campaigns using such means. Fitness apps may analyse smart wearables data recommending workout plans or food supplements based on health journeys of users.

Conversational AI, such as chatbots and voice assistants, is going to become a critical part of targeted marketing in years ahead. These technologies utilize natural language processing (NLP) to hold personalized real-time conversations with customers. Chatbots simulate personal shopping experiences giving immediate suggestions when asked by any individual. In terms of examples, consider a garment store's AI that can support you in choosing outfits according to your taste or past acquisition records which will help improve shopping while increasing sales. Eventually, voice search and smart speakers will soon become an integral part of the digital environment, which is why voice-activated marketing will be on the way up. Brands must create voice-friendly content including advertisements designed for voice search inquiries to stay competitive in this new space. If someone searches "quick dinner ideas" using their voice, a recipe app for example may pay to ensure that its recipes come up first.

7 Social Media Ensures Sustainable Marketing

Apt real-time insights provided by social media platforms reduce inefficiencies and resource wastage that could be experienced in traditional methods of advertising. In the past, the absence of immediate data gathering and evaluation resulted in unrelated campaigns ignoring prevailing customer wants hence lowering productivity as more money was directed to wrong avenues.

7.1 Agility in Marketing

Being able to draw real-time insights from social media will enable marketers to be agile and quick to respond to the needs of their customers. In events like the COVID-19 pandemic, for instance, there have been rapid shifts in consumer preferences due to the influence of social media (Miah et al., 2022). An essential aspect of marketers surviving within a fast changing world is their flexibility. Companies can quickly switch their marketing strategies based on customer feedback or new market conditions because they are able to take advantage of the flexible nature exhibited by social media insights. This highlights the importance of agile marketing practices supported by real-time information which facilitates adoption of more impactful marketing strategies at all times (Chopra et al., 2020). The fast pace at which tastes change in a dynamic environment necessitates such flexibility.

7.2 Reduction of Marketing Waste

One of the key benefits of using social media for consumer behaviour research is in terms of marketing sustainability. In traditional marketing research methods, surveys or focus groups were often collections of answers by only those who felt like responding, ignoring the vast majority opinions. There used to be time lapses between gathering information, analysing it and subsequently conducting campaigns that met customers' desires. This strategy may now seem to have resulted in wrong resource distribution or utilization which hampered sales productivity as well as misallocating funds inappropriately. In contrast, current consumer feelings are depicted by real-time data obtained through social networks. Therefore, it is important for organizations to aligning their spending by allocating it effectively (Baird and Parasnis, 2011), hence ensuring relevance between the two entities mainly marketers' plans and what is happening within a marketplace (Lim et al., 2017).

Social media platforms are now vital accessory tools for facilitating sustainable marketing practices and IKEA is used as an example to demonstrate how businesses can utilize these forums in line with their sustainability goals. Ikea's marketing strategy employs social media analytics which enhances effective communication making it savings-wise while being eco-friendly at the same time. The company is able to track consumer trends and sentiments by monitoring conversations on various networks. Some of the tools for analysing feelings helps identify which factors are most important for consumers and uses these results to change the style of the adverts so that it resonates well with its target audience. Through the knowledge about what is important in social media related to consumers, IKEA has been able to cut down on marketing wastes and at the same time engaging consumers at a deeper level.

8 Social Media Research and Ethical Considerations

While social media research provides countless opportunities, there are however quite a number of ethical issues that need attention first. Among these include infringement of privacy which implicates people's rights, inaccuracy of data therefore wrong decisions or conclusions may be drawn from them, and manipulative marketing strategies.

8.1 Privacy Concerns

Consumer privacy is likely to be breached during social media research and this is a major ethical problem. Sites usually access a lot of personal information such as browsing practices, preferences and geolocations even without explicit permission from users. There are notable ethical concerns regarding fairness and openness in terms of data collection methods.

To this effect, frameworks like the General Data Protection Regulation (GDPR) in the European Union and the California Consumer Privacy Act (CCPA) in the United States, aim to provide people with some control over their information. One of the provisions under GDPR is that entities need to ask for specific permission before they obtain or process anything related to them on the internet. Those parties get lifetime license known as "right to forget" letting them remove the materials if they wish. This is true for CCPA, which provides individuals in the state of California with a right to be informed about who collects their details, why they do so, and if it is shared with other organizations after which one can opt out.

The Cambridge Analytica incident where Facebook was accused of using user's data for political campaigning without their knowledge, illustrates the importance of adhering to the abovementioned privacy legislation in order to prevent data misuse and the erosion of consumer trust. Globalization has made it easier for many companies to track down their customers' personal information that they use to personalize products or services designed specifically for them but this should not come at the expense of their clients' confidentiality.

A way of ensuring this balance is through an opt-in policy on data collection by companies. The opt-in policy seeks to gain approval from a user before any data is collected or analysed unlike the opt-out mechanisms that assume consent. When these strategies are in place, users shall be able to decide on their own about how much of their personal life is publicized. For example, the App Tracking Transparency feature on Apple devices urges customers to proactively choose whether their information may be tracked or not thereby setting higher standards in digital consumer rights. One needs to strike a balance between personalisation and confidentiality and provide clear guidelines on how they intend to collect, process and utilize client details if they want trust from clients.

8.2 Data Accuracy

The accuracy of data collected from social media is one other crucial ethical issue. Social media platforms abound may affect the reliability of consumer insights because there are many spam accounts, fake accounts, and manipulated content. Bots can flood social media feeds with spam or disinformation leading to inaccurate analysis if not properly filtered with anything other than the manual methods (for example using spam traps). To address this issue, companies are forced to put in place a strong validation procedure that will help differentiate between genuine and not genuine user-generated posts. Thus, data accuracy and reliability can be improved through such techniques as AI-powered bot identification, cross-referencing user activity patterns, and incorporating metadata analysis. Social media platform X, has put in place advanced bot-detection algorithms to curb fake news dissemination and enhance quality of user-generated content. By so doing, consumer behaviour research based on social media accurately represent people's real behaviour instead of automated interference.

Another issue revolves around the misinformation and manipulated content available everywhere on social media. A single viral false account or a biased narrative seems to distort the consumer sentiment and, with it, leads to wrong conclusions. What is more dangerous is that due to their training with such siloed data, AI models may adsorb these errors and then propagate them. Developments in AI encompass new algorithms that are suffused with inherent biases, which are effectively replicating existing social prejudices. There is thus concern that bias in models could lead to distinctly biased targeting or personalization where particular groups are purposely excluded or cast in a negative light. In such an environment, without human intervention and strict data quality controls like ethical algorithm design, rigorous evaluation of output, and the continuous winding and rewinding of these systems, good results become unpredictable and complex.

8.3 Manipulative Practices

Human beings can also be exposed to manipulative practices involving media research where highly personalized marketing is used in manipulating their buying habits. This personalization through algorithms may unknowingly bring about 'filter bubbles', which ensure that only contents that match ones already held ideas get through to such a person thus limiting alternative point of views being aired out. The principle of ethical advertising requires an equilibrium between the two elements; human being's choices and personalization by business. In this case for example, GDPR has specific clauses protecting individuals from automated decision making requiring human intervention on grave matters concerning people's lives or futures thereby avoiding situations where programmed decisions enhance particular sellers over others. Firms could make goods or services without involving consumers hence promoting consumer autonomy while utilizing the precision of AI targeting tools; this practise raises concerns relating to manipulation and exploitation of users. To deal with such issues, there should be an integration of ethical audits into the process of creation algorithms in organisations.

Moreover, transparency in advertising practices could eliminate doubts about deceptive advertisements. It is necessary for corporations to explain clearly why certain advertisements are displayed to customers. This can be done through incorporating features like "Why am I seeing this ad?" which can help users comprehend the factors affecting selection of advertisements. The precision that AI driven targeting has may raise questions on manipulation or exploitation. This means they should embrace best practices in order to ensure that the marketing strategies do not compromise consumer's rights but promote favourable experiences instead. This requires adherence to legislative frameworks, investment on advanced validation tools and commitment to transparency as the possible way forward amid these ethical issues that affect business operations. By following General Data Protection Regulations (GDPR) companies that also align their practices with Consumer Privacy Protection Act (CCPA) will not only manage legal liabilities but also build trust among buyers who are now demanding more from firms in such marketplaces dominated by information-driven forces. Furthermore, when these guidelines are implemented in marketing and research strategies then social media may be used carefully so that businesses succeed while ensuring also that customers well-being remains top priority. Below is a general guide on how companies should handle various challenges relating to ethics whenever they are conducting a study using social media networks (Fig. 3).

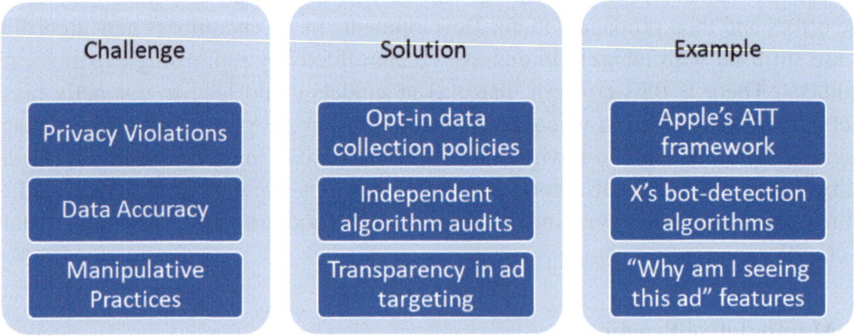

Fig. 3. Practical Steps for Ethical Social Media Research

Marketers should support consumer autonomy and not undertake exploitative measures like hyper-personalized persuasion or excessive nudging. Transparent presentation and explicit opt-in opportunities that shed light on consumer choices will suffice for intelligent regulatory measures such as the EU Digital Services Act where transparency will be a priority moving forward and will make digital governance much more consumer-oriented.

9 Conclusion and Future Perspectives

Consumer behaviour research is likely to be increasingly affected by the convergence of technology, with AI, big data, neuroscience, and behavioural economics interwoven together. This tight interwinding opens up opportunities for extremely fine-grained insights but also fuelling complex ethical questions that academia and industry will need to address collectively.

According to a plethora of literature, it is clear that social media data is increasingly becoming popular for informing marketing campaigns. For this reason, marketers realize that they need to use social media as a strategic resource for better understanding their users' needs and concerns as well as resource allocation that would be more sustainable in times of digital market development. Social media has changed the way companies conduct research on consumer behaviour against what it used to be during the pre-social networking era while conducting business in addition to providing quicker, more reliable and broader insights. Marketing approaches have undergone a revolution due to the fact that companies can now interact with customers in real time thereby making decisions fast and avoiding wastage of resources that would have been used for data collection. Social media analytics enable companies to develop targeted campaigns that resonate with the audience while having efficient operations. But like any other transformative tool, it has its own set of issues including but not limited to manipulation, reliability and privacy concerns.

In the future consumer behaviour research will be more enhanced by AI. Currently AI powered devices, like machine learning algorithms or natural language processing are capable of analysing big data to unearth patterns that were previously hidden; it also predicts future customers' actions quite accurately better than human beings could ever do. Consequently, when these types of inventions undergo more development stages, there is a possibility that they may extend their capacity even further in different applications such as for instance, not just what people are likely to buy but also on which platform they are being swayed by which message and so on.

The key area of growth in personalization is hyper personalization of communication structures by means of singling out specific populations rather than addressing users at large through personalized design process. AI has the potential to comprehend an individual's needs using social media habits, online purchases and browsing activities thus it can create tailored online shopping solutions for each client from product suggestions to personalized adverts. When AI starts to play a more crucial role in consumer behaviour analysis, the moral concerns arising from its application will grow exponentially. These include safekeeping private information, algorithmic prejudice and deformation of clients. As long as AI achieves its goals properly, these problems should be approached as from today. The entire process of data collection and analysis should be transparent, accountable as well as fair.

Ethical marketing using AI needs future scholars to come up with standards. What is the method that businesses can adopt so that they avoid enhancing prejudices through their algorithms while at the same time taking advantage of the weaknesses vulnerable consumers have? What structures should merchants employ ensuring that user information is applied lawfully with definite content from them? This needs probing into by academic institutions as well as representatives from industrial sectors plus regulatory authorities. Research in this area will also depend on interdisciplinary work on customer behaviour into the future. A better understanding of why people do things can come out by putting together findings of psychology, sociology, computer science and marketing. An understanding developed this way will show subtle links between choices made by an individual person at a certain time by considering cultural changes and technological advances over history. Subsequent studies might focus on how new technologies such as AR or VR impact purchasing behaviours or consider leveraging blockchain for greater transparency in data collection and management. Figure 4 shows the progress made in consumer behaviour research over time as well as the key directions that will guide its development.

Future research will have to dig deeper into problems such as how AI tools can be audited for fairness and interpretability, how blockchain can help, keeping the transparency of data in view, and how interdisciplinary frameworks, colluding psychology, data science, and ethics, can push for the paradigms of the future of consumer behaviour research in a manner socially responsible.

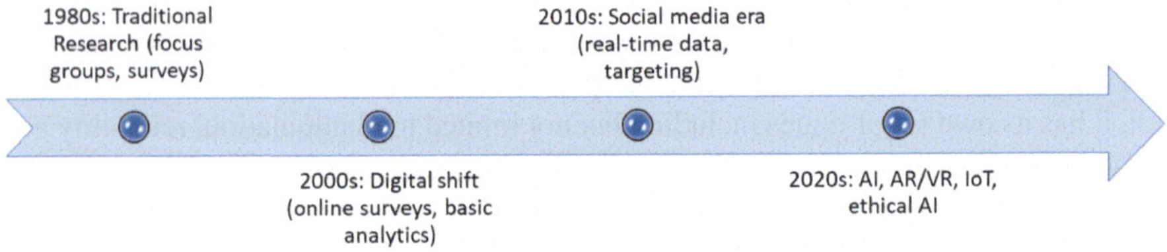

Fig. 4. The Evolution of Consumer Behaviour Research

Continued transformation is on the verge within consumer behaviour areas because there are continued changes in tech advancements together with transforming consumer interests. Social media has changed completely how things were done in this area but AI integration, ethical considerations, and sustainability are what marks where things are headed in research. To tap into all aspects of consumer behaviour insights which can enable both effective and ethically responsible marketing strategies, companies need to observe these emerging trends as well as take up any challenges posed by them. In taking this path organizations should also keep in mind not only innovation but equally important the human element of consumer behaviour in their efforts to innovate forward. For corporations which traverse across these changing times, research on consumer behaviour does not only lie on science and technology advancements but also in its capacity to create meaningful associations with customers that are meaningful and ethical.

References

Chopra, A., Avhad, V., Jaju, S.: Influencer marketing: an exploratory study to identify antecedents of consumer behavior of millennial. Bus. Perspect. Res. **9**(1), 77–91 (2020)

Godey, B., et al.: Social media marketing efforts of luxury brands: influence on brand equity and consumer behavior. J. Bus. Res. **69**(12), 5833–5841 (2016)

Heller Baird, C., Parasnis, G.: From social media to social customer relationship management. Strategy Leadersh. **39**(5), 30–37 (2011)

Koesharijadi, K., Merthayasa, A., Nendi, I.: Development of digital marketing strategy and service quality using social media towards digital economy. J. Soc. Sci. **3**(4), 799–804 (2022)

Kumar, A., Bezawada, R., Rishika, R., Janakiraman, R., Kannan, P.: From social to sale: the effects of firm-generated content in social media on customer behavior. J. Mark. **80**(1), 7–25 (2016)

Li, F., Larimo, J., Leonidou, L.C.: Social media marketing strategy: definition, conceptualization, taxonomy, validation, and future agenda. J. Acad. Mark. Sci. **49**(1), 51–70 (2020)

Lim, X., Radzol, A.R.B.M., Cheah, J., Wong, M.W.: The impact of social media influencers on purchase intention and the mediation effect of customer attitude. Asian J. Bus. Res. **7**(2) (2017)

Mason, A.N., Brown, M., Mason, K., Narcum, J.: Pandemic effects on social media marketing behaviors in India. Cogent Bus. Manage. **8**(1) (2021)

Miah, M.R., Shikder, R., Hossain, A., Saha, T., Neger, M.: Evaluating the impact of social media on online shopping behavior during covid-19 pandemic: a Bangladeshi consumers' perspectives. Heliyon **8**(9), e10600 (2022)

Liu, P.: Consumer behavior prediction and market application exploration based on social network data analysis. J. Electr. Syst. **20**(6s), 806–811 (2024)

Rahmani, S., Kamila, E.R., Violita, C.E.: Nfts and branding: a marketer's guide to the digital art craze. J. Econ. Bus. Account. (COSTING) **7**(4), 7914–7919 (2024)

Yadav, M., Rahman, Z.: The influence of social media marketing activities on customer loyalty. Benchmarking Int. J. **25**(9), 3882–3905 (2018)

The Impact of Virtual Try-On Tools for Beauty Products on Consumer Behavior: A Segmentation Study

Anna Zarkada[1], Suzanne Barakat[2](✉), and Yioula Melanthiou[1]

[1] Cyprus University of Technology, Limassol, Cyprus
anna.zarkada@cut.a.cy, yioula.melanthiou@cut.ac.cy
[2] University of Salford, Salford, UK
suzanne.barakat@yahoo.com

Abstract. We present the characteristics of Virtual try-on tools (VTOs) with an emphasis on digital mirrors used by the beauty industry and review the extant knowledge on the mechanism by which they influence consumer behavior. We utilize a quasi-experimental empirical research design through which 221 self-identified as female participants tested the VTO of a luxury makeup brand and then completed a survey of their socio-demographics, attitudes, and behaviors. Two segments emerge, the hedonic enthusiasts and the utility-focused consumers. Their characteristics are analyzed, and the insights are utilized to draw recommendations for marketers in the beauty industry. This is one of the first studies of how VTOs influence consumer behavior in the beauty industry and the first attempt at segmenting their users.

Keywords: Virtual try-ons · digital mirror · augmented reality · market segmentation · consumer behavior · beauty products

1 Introduction – The Beauty Industry Going Digital

The global beauty industry, comprising a wide range of products under the broad retail categories of makeup, skin care, haircare, and perfume, is a fast growing one. It generated over US$ 400 billion in 2022 [1], grew by 10% in 2023 [2] and is expected to reach almost US$ 600 billion in 2027 [1]. Part of the growth is attributed to prices rising in line with inflation in key markets such as the EU and USA as well as the introduction of technologically advanced and therefore pricey products. Most of it, however, is the result of volume growth. For example, in the skin care category, growth is driven by emerging consumer segments, such as Gen Z, Alpha Generation, and men of all generational groups, as well as product innovations, like the " buzzy ingredients" [2], neuropeptides, which are small protein-like molecules acting as messengers to the cells to produce collagen or triggering skin firming and revitalizing processes [3]. The haircare category is also estimated to have a compound annual growth rate (CAGR) of 5% in the 2023–2028 period due to global trends such as the "skinification of hair, the introduction of

A. K. Zarkada (Ed.): GMA-GAMMA 2025, SPBE, pp. 104–121, 2025.
https://doi.org/10.1007/978-3-031-96930-0_7

elaborate routines and specialized products to the hair category, as has been the case in skin care" [2]. Finally, despite the anemic growth of consumer disposable income in the USA [4] and EU [5] consumers are still splurging on beauty products; with some segments such as Gen Zers in China and millennials in the UAE, or makeup in the Asia–Pacific, Middle East and Africa regions emerging as big industry growth drivers [2].

Like many other industries, retail beauty products have found their way to the digital domain. The e-commerce channel achieved a roughly fourfold increase since 2015 [1], helped the industry survive the COVID-19 lockdowns [6], exceeded 20% of the total industry size by 2023 [1] to emerge as "the largest single channel for beauty globally" [2], and has remained the fastest growing one with a 2023–2028 CAGR of 12%. To put the latter point in perspective, the respective figures for travel retail spaces (e.g. airport duty free), specialty beauty shops, and department stores are 9%, 8% and 2% [2]. Part of the channel's growth is a matter of accelerated investment in infrastructure in large markets such as the Middle East and Latin America, increased mobile-device usage, and worldwide improvements in logistics and payment systems. Other contributing factors are the move of established direct-to-consumer brands to e-commerce, the increased sophistication as well as increasing numbers of omnichannel retailers, the explosion of livestreaming and other social selling practices, and the expanding portfolio of online marts like Amazon and its Chinese counterpart, Tmall [1].

This, however, is not an indicator of the online channel cannibalizing the rest. Consumer behavior is both ROPO and reverse ROPO with about half of the consumers in some of the biggest markets like Brazil, Europe, Japan, and the United States exercising the traditional research online purchase offline, about a third in all countries researching only online, and about half researching only offline and purchasing offor online [2, 7]. The majority of consumers in established markets, such as France, Germany, Italy, the United Kingdom, and the United States, visit both online and offline stores to try new brands [1] whilst in emerging markets, such as Vietnam, young females prefer to have their beauty products delivered [8]. The customer journey is clearly, and globally, a phygital one. Customers seamlessly move between digital and physical channels and craft personal omnichannel experiences which form the basis of building relationships with new brands whilst sustaining their relationships with legacy brands. There is a good reason behind the increasing complexity of the retail beauty industry's phygital customer journey: consumers value the convenience of e-commerce [8] but they are not willing to pay more for it, so they shop across price points [1]. Most importantly, they need to try and test products before deciding to purchase [9].

2 Virtual Try-On Tools

Fundamental to the development of the online beauty market is the introduction of Virtual Try-On tools (VTOs) which overcome the problem of consumers' need to try and test products before deciding to purchase. The VTOs utilize Artificial Intelligence (AI) and Augmented Reality (AR) to allow consumers to try various products such as haircoloring and makeup in real-time on themselves using their computer or mobile phone screen. They create an immersive shopping experience for consumers and open new communication avenues for businesses. Their smart AI analyzes the user's face captured by their smart device camera and then places the selected product right over the area the user has selected. To a great extent, VTOs' immersive reality closely resembles shopping in the physical world. In the case of make-up, for example, it reduces the need to visit a physical store and eliminates being touched or convinced by salesperson's. In other cases, such as with boxed haircoloring, the augmented reality overcomes the limitations of choosing the right product in a physical setting which can only be done by gross approximation – the tiny snippets of how the color looks on light or dark hair found on the side of the box. In general, using VTOs, customers can easily try products from e-shops that can be found anywhere around the globe and make an informed shopping decision. In this sense, this technology helps businesses build customer trust and supports brand loyalty.

Initially, VTOs were introduced in niche retail industries like virtual fitting rooms (VFR) in fashion sector [10]. Early research showed that having a VFR in a website not only increased interest in a specific product but also affected purchase intent and propensity to patronize the e-shop as well as the physical store [10]. This experimental research result supports our assessment, above, that the digital and physical channels are complementary rather than the new one cannibalizing the old ones. During both ROPO and reverse ROPO, consumers use AR technology, alongside product information and user generated content (UGC) in the form of videos, images, and text, as bridges between the digital and the physical world. As recent survey research has demonstrated, users who have become familiar with AR technologies go past the initial interest towards it, which was a reflection of its novelty [11], vividness [11–13], interactivity [11–13] and playfulness, and engage with it to exploit its augmentation potential [13, 14] to meet their utilitarian objectives [6, 14]. Eventually, the use of AR tools increases purchase intent [11, 12, 14]. Thus, the rise of revolutionary interactive technologies such as AR, which merges images of reality with virtually added content to create a uniform user experience, have, not just disrupted [15], but, fundamentally changed the retail industry dynamics regarding how consumers interact with products and businesses online [16], by bridging the sensory gap between eand m-commerce on the one hand and brick-and-mortar channels on the other [17]. Of all the AR technologies used by marketing the VTOs enhance the functionality applications and websites by facilitating consumer decision-making [9], addressing hygiene and safety concerns18 and reducing post-purchase cognitive dissonance, and in this sense contributing to addressing the returns challenges for businesses [19].

2.1 Digital Mirrors

The academic work reflects the realization of the potential of using VTOs by the beauty industry. The digital mirror is the newest VTO tool which is commonly used in the beauty industry to allow consumers to virtually experiment with options of shapes, textures, colors, and shades on their own face, hair or nails. Technically, digital mirrors (also referred to as AR magic mirrors or virtual mirrors), like all VTOs, belong to the self-augmentation category of AR technology [20] (the other two being augmentation of the environment and object augmentation) as they produce the users' altered perceptions of their appearance through the beauty VTOs [49]. They are a more realistic and interactive version of earlier technologies, such as websites in which consumers could create their avatar on which they could try clothes. In this sense contemporary digital mirrors are an experiential marketing tool [21] which takes the consumer's imagination and engagement ot whole new level of engagement.

In practice, digital mirrors are key e-retailing tools of makeup essentials, like eyeshadow, lipstick, and foundation, but also haircoloring or nail polish, [22]. Since buying makeup depends on maintaining a high level of visual involvement [15], VTOs, such as the digital mirror, substantially enhance the appeal and functionality of corporate digital media and this has led to their extensive adoption by industry players [23].

Sephora™ and Ulta Beauty™ launched their digital mirrors in 2016, known as Virtual Artist and GLAMLab [24]. In 2018, L'Oréal™ incorporated VTO tools into different digital platforms to differentiate itself from competitors [25]. This move resulted in e-sales increasing by 52%, so, in 2021 the personal care giant launched a social VTO in partnership with Facebook [26]. The luxury cosmetics brand Charlotte Tilbury™, one of the first brands to add a VTO on a website [27], implemented the cutting-edge "Magic Mirror Makeup Online Try On" after the COVID-19 pandemic [28]. So, not only new entrants to the beauty market, such as Glossier™ and Rihanna's Fenty Beauty™, but also major international brands, such as Revlon™ and MAC™ have, over the last decade, increasingly been incorporating digital mirrors into their digital platforms for their haircoloring, makeup, and nail polish product lines.

Consumers have been found to appreciate digital mirrors for their convenience, fun, and user-friendliness [22]. The positive emotions they evoke have been found to facilitate product choice and positively impact purchase decisions [9] by cocreating bespoke shopping experiences. So, it comes as no surprise that the global virtual makeup try-on market size is forecasted to grow by a CAGR of 23.3% from 2024 to 2033 [29].

3 Understanding the Impact of VTOs on Consumer Behavior

Academic researchers examined AR-powered applications (apps) like VTOs and digital mirrors through various disciplinary lenses forming an expanding body of literature documenting the development of this cutting-edge e-retail technology [30], its functionality and affordances as a communication medium [20] and its effects on consumer behavior [21, 26, 31] but also on their self-concept [32]. As most of the extant literature has examined a variety of AR technologies, in this section we present their conceptualizations and findings adjusted to the particularities of VTOs, and digital mirrors in particular.

3.1 Characteristics of VTOs

Analyses of AR technologies as media [11, 20, 33] identified the following characteristics of smart device-based VTOs:

1. High ***Interactivity*** between person and machine allowing consumers to actively engage with three-dimensional virtual models of products being used. The high degree of interactivity turns using the app into a fun and pleasurable experience, but also a safe one as the user is in full control of the immersion experience.

2. High ***Vividness*** as the resulting image is as close to actually applying the product and looking at oneself in the mirror as technology can afford. Contemporary digital mirrors produce images which are as vivid as reality but without the mistakes a user can make in application of the products.
3. Diminishing ***Novelty*** as the more widespread the use of VTOs becomes, the more consumers become familiar with them to the final point of taking them for granted and seeking other options to stimulate their imagination.
4. High ***Informativeness*** as they offer useful and totally relevant information about specific products which help consumers make fully informed decisions and thus reduce perceived risk, especially functional risk, and subsequently limit post-purchase cognitive dissonance. Interactivity, vividness, novelty, and informativeness, combined with measures of the functionality of VTOs encapsulated in the concept of ***System Quality*** have been used as the dimensions of the stimulus in S-O-R theory-based empirical research studies [11].
5. ***Hypertextuality*** with many sources linked into creating a phygital experience comprising a blended reality with actual and projected, imaginary elements.
6. Multi-***Modality*** of content representation with textual, visual, and audiovisual components. Hypertextuality and Modality are combined to create the ***Augmentation*** which involves overlaying virtual elements onto the real world, adding a layer of novelty and enjoyment to the shopping process, and delivering further entertainment and immersion
7. One-to-one ***Connectivity*** – that is, a totally personalized experience - with the potential to share the images on other media.
8. Medium to high ***Location specificity*** controlled by the users who can allow targeting using geolocation or limit it to manually choosing the region so that they can see the products they have access to and manually inputting shipping and payment details.
9. Total ***Mobility*** as digital mirror apps on mobile devices can be easily carried and used anywhere.
10. Medium ***Virtuality*** as the virtual elements are immersed in images of reality but do not create a totally virtual experience isolated from reality (i.e. the user's face or hand).
11. High ***Ease of Use*** as their use is simple and straightforward and unlikely to cause consumer frustration, even to novice users. This, of course, is an evolving characteristic; the more technology advances, the easier it becomes to use but also the more adept the consumers become in using it.
12. Limited to no ***Telepresence*** as digital mirrors, unlike other X-realities do not attempt to create a sense of being physically present in a virtual environment. They usually make the consumers' private environments part of the shopping and product use experience. In this sense, they situate the product in the consumer's reality. Thus, it is the product that is being immersed and enmeshed in the consumer's frames of reference which makes the experience feel realistic and enhances both enjoyment and efficiency.

In the context of beauty products, these characteristics of VTOs play a role in delivering engaging and efficient experiences. Informativeness and ease of use boost utilitarian value by offering detailed product information and user-friendly interfaces that help reduce uncertainty and simplify product choice. For makeup products, this means providing shade details and seamless browsing through various items or combining different products together to create a complete 'look' so they promote not only sales but also cross- and upselling.

Interactivity, augmentation, and telepresence are key factors enhancing the hedonic side of value perception by creating immersive experiences that allow consumers to visualize how different products would appear on their skin tone and facial structure, thus elevating the realism of the VTO experience. It is worth mentioning that out of all these characteristics, augmentation stands out for its ability to provide modality richness by stimulating multiple senses with greater intensity than static pictures or text, thereby creating a deeper sensory experience that feels not only engaging but also realistic and fully personalized. Fundamentally, augmentation ensures the integration of physical reality with virtual content, crafting a unique, authentic user experience.

The self-augmentation VTOs afford has been empirically found to enhance self-brand connections and purchase intentions, especially for narcissistic consumers [34]. This finding suggests that consumers perceive higher experiential value when they see themselves using the product rather than viewing professional models. There is also empirical evidence of the positive effects of informativeness, ease of use, and telepresence on consumers' utilitarian value perception and of telepresence and interactivity on hedonic value perception [33]. During the past decade, there have also been some empirical research results which have demonstrated that the particular combination of the characteristics of VTOs impact on consumer perceptions, including self-perception [32], cognative [21], affective [21] and behavioral [21, 33] responses [21], and behavioral intents such as purchase and patronage [6, 10, 13, 14, 26, 34].

In summary, a 2022 systematic literature review of 53 articles on the use of AR in e-retail [35] concluded that "AR characteristics significantly influence utilitarian, hedonic, perceived risk and experiential value, ultimately resulting in a positive attitude, decision-making assistance and behavioural intentions, wherein customer experience (flow, spatial presence, mental imagery and immersion) plays a mediating role in the process". Augmentation was found to be the most influential in terms of behavioral intentions. The drivers of the behavior were identified to be the perceived value in both the hedonic and utilitarian dimensions whilst perceived risk was, as expected, the most influential negative influence on behavioral intent. The following dimensions of perceived risk were identified: perceived intrusiveness, privacy concern, perceived risk, and control over information access. The results regarding the links between positive attitude, purchase intention, and e-word of mouth (e-WOM) were mixed – a common finding of systematic literature reviews and metanalyses in marketing.

3.2 How VTOs Impact Consumer Behavior

Academic explorations of the mechanism through which VTOs influence consumer behavior in retail and e-retail settings started after 2014 and have mostly applied classic theories such as TAM, "flow", situated cognition theory, and equity theory [19, 35]. Indeed, in our literature review, most empirical studies were based on the Technology Acceptance Model (TAM) [6, 17, 23, 31, 36, 37] and its variants such as the modified e-TAM [17] or the Value-based Adoption Model (VAM) [26] whilst some used the

Stimulus Organism Reponses (S-O-R) model [9, 11, 33, 38] and mixed approaches combining S-O-R logic with elements from multiple theories in new configurations [19] or the Uses and Gratification Theory (U&G) [39]. It is beyond the scope of this chapter to explore the details of the theories or to present a comparative review of them. Instead, we have identified the key constructs that influence consumer behavior in the context of using VTOs and the relationships between them as identified in the relevant empirical literature.

Attitude and Perceived Value

Consumers with positive attitudes towards a new technology like VTOs are more likely to adopt it if they perceive that they derive a certain value from using it. Perceived value in the digital landscape is the result of the consumer's processing the experiential rather than the expected use or consumption component [38] It is a bi-dimensional psychological construct; it can be experienced as hedonic and utilitarian value [9]. Hedonic value is the positive outcome of multisensory perceptions of enjoyment, fun, or socialization derived from an experience, and it is mostly affectively oriented, whereas utilitarian value is associated with cognitive and practical aspects in acquiring useful information or achieving functional benefits from an online service. This is not to imply that these two components of perceived value are independent of each other. They are interrelated influences on the purchase process despite the fact that the consumer can distinguish between (a) the utilitarian value derived from the degree to which VTO app functionality, usefulness, and ease of use balance the perception of transaction and privacy risks involved in using the app and (b) hedonic ones such as fun, socialization, and pleasure [39].

Perceived hedonic value has been empirically found to be a predictor of the intention to continue using VTOs and intention to purchase the tried-on products [11, 33] whilst in other studies, higher perceived utilitarian value was found to be related to purchase decisions [39] and intention [33]. The results are not conclusive as the study samples [39] are heavily biased towards the younger generations and people who have been primed or have prior experience with VTOs. Moreover, it is not yet clear how the context and product dimensions interact with the age of the consumers and their attitude towards technology in general.

Anyway, when the perceived value meets consumer value expectations from the VTO it drives customer satisfaction. Customer satisfaction increases positive attitude towards both the VTO tool and the product, which in turn predicts consumer's behavioral intentions including channel patronage [40]. It has to be stressed that marketers in the beauty sector routinely prioritize the cocreation of experiential value over a traditional transactional approach (Hudson et al., 2018). So, VTOs serve as an online evaluative tool aimed at enhancing the experiential component of search stage of the consumption decision process. The more engaging the experience, the more the consumer consideration set is narrowed, and the more likely the product purchase and repurchase becomes.

Characteristics of VTOs, Attitudes and Perceived Value

Augmentation aims to combine the hedonic and utilitarian dimensions of value to enrich user experiences and produce a positive affective response. It also shapes the brand consumer communication framework through which behavioral intentions [21] and actual behaviors [9] are formed. The novelty, immersion, enjoyment, and usefulness of VTOs have been found to be related to positive attitudes towards the app itself as well as higher levels of product purchase intention than simple digital product presentations [41]. Similar results have been reported in recent quasi experimental studies [37] which found ease of use, usefulness and enjoyment of using VTOs to be positively associated with attitude towards the app and, indirectly, with behavioral intention.

Moreover, field experiments comparing an AR app and a website offering a unique visual illustration of the product attributes showed that an immersive experience was perceived as having higher hedonic value compared to websites, leading to higher customer satisfaction which in turn translated into positive attitudes. Indeed, AR interactivity and vividness influenced both perceived usefulness and enjoyment which means that it offered consumers higher hedonic and utilitarian value, that is its total perceived value was higher than that of the static images of the website [41]. Even in early AR apps, which were much less vivid and functional than the ones available today, the augmentation effect boosted emotional responses such as excitement, arousal, and pleasure, which consequently enhanced the overall shopping experience, influenced consumer perceptions, and facilitated decision-making. Finally, the greater consumers perceive medium usefulness and enjoyment the more positive their attitude towards VTOs becomes [41]. Augmentation has been found to positively influence hedonic value perception [33], and to generate strong affective responses, and through them significantly influences purchase intention, especially for consumers driven by a hedonic shopping motivation (Watson et al., 2018). Interactivity has also been found to positively affect hedonic value perception [33].

Finally, interactivity and novelty have empirically been found to influence hedonic value perceptions which in turn influence satisfaction and consumer intent to continue using the VTO which is the final step to purchase intention in the decision process [11]. The same study [11] showed that hedonic value perception is not affected by the perceived vividness, informativeness, and system quality of the VTO tool. In another study [33], informativeness has been found to positively affect utilitarian value perception as is ease of use and informativeness.

Despite the growing body of research about the impact of VTOs on consumer value perceptions, attitudes towards them and the products they promote, and behaviors, the factors that may intervene in these relationships are still underexplored. Key issues in need of further investigation are the influence of the consumers' social environment and their personal traits and characteristics, both of which are underexplored [35].

Subjective Norms

Subjective norms can be likened to society's unwritten rules that influence what people think they should or should not do and, in this sense, they measure the social pressure a person feels from important others, such as family, friends, peers, or aspirational groups, to act in a certain way [42]. Understanding consumers' subjective norms helps predict how inclined they would be to use VTOs based on a recommendation from a close friend or preferred beauty influencer. It would also predict the degree to which the social aspect of utilizing VTOs when shopping for beauty products, which requires consumers to share their images or get feedback from their 'purchase pals', would enhance perceived social validation which leads to better choices and increased interest in products [43] or if the sharing would be perceived as an act of oversharing or non-permitted level of exposure of personal information to the degree that it would stop the consumers form using the VTOs.

When the images they produced are shared on social media, VTOs facilitate seeking peer opinions at the pre-purchase stage which reduces the level of perceived social risk. Positive feedback from peers was found to be associated with higher purchase intention among consumers. Interestingly, participants who received negative feedback from peers but had exposure to a VTO tool reported increased purchase intentions compared to those without VTO exposure. This suggests that social endorsement can influence consumers' purchase intentions, especially when combined with AR experiences [12].

Finally, there is some evidence that, contrary to what most studies have shown, using VTOs may lead to negative emotions such as low self-esteem and they can amplify impulse buying propensity which may result in feelings of guilt (Yim et al., 2017). Fear of being perceived as vain or too materialistic might emerge as a perceived social and psychological risk factor which could discourage some consumers from reusing VTOs in their online shopping for cosmetics. Moreover, VTOs have been found to influence the way consumers perceive themselves. AR mirrors can, under certain circumstances, widen the gap between consumers' ideal appearance and their perception of their actual appearance. They can also affect product-related variety seeking behaviors which can be related to benign aspects of self-evaluations, like consumer confidence in the look the VTO proposes, and even sensitive ones such as mental well-being and self-compassion [32]. Thus, individual differences emerge as an important empirical exploration topic.

Psychological Traits and Sociodemographic Factors

There is some isolated and rather inconclusive evidence regarding the effect of specific personality traits, such as narcissism [34], appearance self-esteem [32], sensation-seeking tendency, and technology anxiety [38] on attitudes towards VTOs and their subsequent effect on consumer behavior.

When using VTOs, consumer characteristics such as age, gender and ethnicity, and traits, such as shopping orientation and product involvement, have been found to indirectly affect purchase intention, satisfaction, and loyalty [9]. In other studies age and gender were found to have no effect on the purchase decision process [12, 39].

The literature we reviewed for this chapter has not directly investigated the influence of socioeconomic status, using proxy measures such as income and education, on consumer value perceptions or attitudes towards the experiential or transactional usefulness of VTOs. More often than not, the samples in empirical studies are biased towards higher educational status [37] and this also restricts the metanalysis potential. In a study of Saudi consumers using AR apps for apparel shopping [36] sociodemographics were found to have some effect on attitudes towards using the technology with women and married people being more likely to hold a positive attitude but age or household size having no effect. Additionally, well-educated females with high incomes showed a stronger inclination towards hedonic value, in their buying intentions [36]. Nevertheless, as with any study utilizing narrowly localized data, caution must be applied, as the findings might not be directly generalizable.

The distinction between utilitarian and hedonic motives could be the determining factor for segmenting consumers using VTOs for beauty products. Women being the majority of beauty products consumers worldwide, it could be hypothesized that VTOs would appeal to mid-to high-income, educated, working women who are likely to face the demands of busy professional and personal lives and so, would be expected to value convenience and efficiency, over visiting physical stores. It could also be expected that VTOs would also appeal to young tech enthusiasts seeking the entertainment of social sharing of alternative appearances of themselves with like-minded consumers. To test these broadly hypothesized conceptualizations, we conducted an empirical study, key parts of which we present here.

4 An Empirical Investigation of the Impact of VTOs for Beauty Products on Consumer Berhaviour

4.1 Methodology

The empirical study we present here utilized a quasi-experimental cross sectional research design as recommended by earlier research [44]. It departs from the norms of earlier studies in the AR as a marketing tool field by surveying real beauty products consumers who are active on social media after they had experimented with the technology in a real-life, rather than laboratory setting. In this section we present an overview of the study and in the next sections we discuss some of the key findings and their practical implications.

The Participants

With a convenience sample was utilized for the empirical part of our study of the impact of VTOs for beauty products on purchase intentions. Most studies on consumer behavior in the digital domain use student samples [20] despite the well-documented limitations of such samples [34]. Contrary to the many of the studies we reviewed which utilized student samples [9, 12, 17, 19, 21, 23, 32, 34, 38] or student and graduate samples [13, 41] we departed from the norm and approached consumers who were subscribed to beauty-related social media groups, as some earlier studies have done [6, 9, 14, 26, 33]. We sent them in-app instant messages to verify that (a) they identified as female, (b) were over 18 years old, (c) were willing to participate in the study, (d) they were consumers of makeup products, and (e) they were planning to use a mobile phone equipped with a camera to participate in the research in order to control for biases that could be caused by device functionalities and response to AR software.

The VTO Tool

In line with earlier studies [9, 12, 13, 26, 45], we used an existing, and very successful, digital mirror in the luxury makeup segment for the quasi-experimental part of the study. Participants were asked to try an existing VTO tool which is embedded in the website of Charlotte Tilbury™.

Charlotte Tilbury™ is a leading beauty brand founded in 2013 by the renowned British makeup artist Charlotte Tilbury. The brand offers a range of makeup and skincare products that have achieved record-breaking launches. It is an influential brand in the global beauty industry with presence in over 67 countries and flagship stores in major cities, in addition to its digital presence through its website, social media platforms, and mobile applications. The founder's goal is to create beauty products that the world really needs and wants, focusing on making everyone feel beautiful, safe, and confident. Her aim is to reach women worldwide through an omnichannel business model.

The brand has a strong digital presence and interactive technologies like AR mirrors in stores and an online VTO tool called "Magic Mirror Makeup Online Try On" powered by facial scanning, completely digitalizing consumer experience [28]. Their VTO tool allows consumers to try a wide range of makeup including foundation, blusher, eyeshadow, eyeliner, mascara, lipliner, lipstick, and contouring powder. Users can then save looks to compare them or share them with others, making the shopping experience fun and engaging [27]. Their try-on tool is user-friendly, and a video tutorial on their website provides clear instructions on how to use it appropriately.

Data Collection Procedures

To reduce memory bias, the participants were requested to visit the website via a weblink (https://www.charlottetilbury.com/uk/products/charlottes-virtual-online-makeup-try-on) which took them directly to the VTO tool. They were asked to select a product in which they were interested, then click on it and choose the 'try on option'. After they were done testing the product using the VTO tool, they were instructed to come back and share their experience through an online survey on Google Forms™. This approach echoes the approach of earlier studies which used existing virtual mirrors for fashion accessories [34, 41], cosmetics [31, 37], hair color [19, 26], and make up [9, 19, 32].

Measures and Data Analysis

The survey measured basic socio-demographics (age, income bracket, and education), VTO augmentation [21] perceived value comprising the hedonic and utilitarian dimensions [33], attitude and subjective norms [18], and purchase intention [39]. All scales were based on a five-point Likert-type scale with only the ends defined as strongly agree to strongly disagree [46]. All questions were marked as mandatory to control for non-response errors [47]. The data were analyzed using RStudio integrated development environment (IDE) Desktop version, the most popular R programming tool for statistical computing and graphics, chosen for its capabilities in handling complex models and generating detailed insights from the data.

Ethics

Following research norms, the social groups' administrators were contacted via direct messaging and email requesting approval to share the survey with their subscribers while respecting the group policy and community guidelines. A participation sheet was sent to each administrator to brief self-identified females' respondents on the study's topic, aim, procedures, expectations, and their right to withdraw at any point without repercussion. The survey was sent as a weblink, allowing administrators to publish it or permit the researcher to post it on the corresponding platform. A consent page was included at the beginning of the survey, to obtain a voluntary participation agreement. The page clearly stated that the data collected will be confidential and anonymous to protect respondents' privacy. Additionally, the researcher verified that no harm is caused to any party involved in the research. The researcher also ensured that all weblinks were safe and functional and tested the brands website using WAVE tool to confirm that the content is accessible to people with disabilities [48]. Participants were also provided with contact information to ask questions or express concerns about the study.

5 Empirical Findings and Recommendations to Industry

A total of 221 usable responses were collected. The majority of respondents were Millennials (52%) and GenZ (43%), educated to bachelor's (52%) and postgraduate (26%) level, and roughly equally split between the lower (45%) and higher (39%) income brackets. The respondents fell neatly into two clusters: the hedonic enthusiasts and utility-focused consumers. We discuss these in the following sections.

5.1 Hedonic Enthusiasts

The findings revealed a group of females who highly value the hedonic aspect of VTOs. They enjoy the shopping experience itself for the excitement and the fun they derive from virtually browsing through products and trying them on. This preference was found in both digital native cohorts (Gen Z and younger Millennial consumers). Their strong appreciation for online engagement aligns with the tech-savvy nature of younger consumers, who reported that they are likely to seamlessly incorporate VTOs into their makeup shopping routine as they see it as the fun part of the purchase process.

Even though this segment includes women from all educational levels it is skewed towards the middle-income (US$1,000-$3,000 personal income per month). What characterizes this segment is that their purchase intentions towards beauty products that they have tried using VTOs are influenced by their peer opinions and social trends more than any other factor.

This segment of beauty shoppers is more likely to enjoy and feel satisfied with apps that offer visually pleasing experiences than any other conventional product presentation method. Engaging digital content emphasizing the recreational dimension of VTOs for beauty products is what effectively captures their interest and boosts the hedonic value perception they seek. Moreover, being digital natives and used to living a blended offline and online life, they are discerning users of technology and they are not easily impressed by digital gimmicks. Thus, their attitude towards both the app and the products it promotes is highly dependent on the quality of the augmentation and the level of enhancement the VTO affords. The technological sophistication of the VTO is also related to the experience of pleasure they derive from it and this is also a factor on which their positive attitude depends.

Catering for the Hedonic Enthusiasts Segment

Increasingly more beauty products manufacturers but also a lot of retailers are realizing the importance of incorporating the entertainment dimension into their marketing communications. For the hedonic enthusiasts segment, however, this is particularly pertinent but not enough. Constantly improving their advanced AR technologies to create realistic and engaging virtual environments that keep consumers satisfied thus emerges as a necessary and sufficient condition of engaging the hedonic enthusiasts and keeping them engaged and coming back to the digital retail space. Earlier research [33] has demonstrated that the technologically sophisticated dimensions of the VTOs, such as interactivity and telepresence, positively affect users' hedonic value perception. Our work confirms this finding and highlights its importance as it emerges as a beauty products market segmentation criterion.

Moreover, our study revealed that what motivates the hedonic enthusiasts segment to moving from having fun with the VTO to purchasing the products they tried is the opinion of their friends which they seek through sharing their saved images on social media. This means that incorporating customizable light filters in the VTO could not only increase this segment's engagement, as they can see how the makeup they have created would look in different lighting conditions, such as night, sunset, day, etc., but also it would allow them to enhance their photos and thus increase their motivation to share them online. This socializing through sharing makeup looks can potentially result in increasing traffic on the retailer's website.

Another strategy that has been identified as having a significant impact on Gen Z and Millennials' consumer behavior is gamification [49]. Beauty retailers can ride this wave while it lasts by running makeup look contests using their VTOs. Such virtual events can generate not only publicity and social media buzz, which increase traffic to the site and app, but also increase trust in the retailer and the products through usergenerated content (UGC) and positive electronic word of mouth (e-WOM). Gamification and other tactics that encourage social sharing of UGC content increase social media engagement with the brand, enhance the credibility of the VTO tool, and increase social empowerment thus leading to positive transactional outcomes [50, 51].

Hedonic enthusiasts' purchasing behavior was also found to be influenced by social trends. This can be exploited by collaborating with celebrities who not only endorse the use of VTOs but also share their explorations with the app on their own and on corporate social media and interacting with users. The choice of celebrities to partner with requires a systematic exploration of their overall popularity and their following, and engagement levels of hedonic enthusiasts with them. This can add to the appeal of the VTO tool and become a competitive advantage basis for the brand.

VTOs can take a central place in brand communication campaigns aiming at creating shareable experiences that align with the networking lifestyle of the hedonic enthusiasts segment [43]. And what would be more shareable than consumers using the VTO to try on trending makeup looks, or looks created by or for celebrities and influencers? The integration of platforms and technologies in order to make the consumer feel central to the brand communications can only be limited by the marketers' imagination. Given the characteristics of the hedonic enthusiasts, this is key to the full exploitation of the potential of VTOs for the beauty industry.

5.2 Utility-Focused

Our study revealed that is a clearly utility-focused segment consisting primarily of millennials in the highest income bracket (over US$3,000 personal income per month). The key characteristic of this segment is that they share a pragmatic approach to technology use. They showed an interest in the practical aspects of using VTO tools to buy online beauty products and valued usefulness and convenience more than the fun and playfulness of using apps. This beauty products consumer segment is motivated to use interactive technologies by the functionality they offer by combining accurate visualization with detailed product information. This finding is consistent with those of earlier studies which identified informativeness, ease of use, and telepresence as the factors which positively affected consumers' utilitarian value perception [33].

Attracting and Engaging the Utility Focused Segment
As earlier research has indicated, our research confirms that to attract and convert the utility-focused segment into buyers, app developers must design a user-friendly activation process of the VTOs to make them appealing to a segment of consumers who do not necessarily have specific technical knowledge and therefore may be discouraged form using the app if they have to deal with complicated setup issues during their shopping journey [52]. Similarly, marketers could further enhance this journey with video tutorials on VTO use published on their website or social media platforms [33]. The main goal would be to instruct users step by step on how to smoothly engage with the VTO which would also be communicating messages about the usefulness of the tool and the ways in which it could help them virtually explore product details alongside trying the products on themselves thus facilitating decision-making without the need to visit stores.

Earlier research has also indicated that purchase intention is formed by affective and cognitive responses to the characteristics of AR [53]. Our research findings concur. However, utility-focused consumers are motivated mostly by evaluations of the usefulness of the VTOs based on system interactivity and augmentation quality, and above all product informativeness and reality congruence. Incorporating into the app features such as customer reviews, ratings, and discussion forums where consumers can chat about the VTOs and the products they have tested virtually or bought can increase perceived usefulness of the app. This can make the VTO and, by extension, the products, particularly attractive to utility-focused consumers (Hong & Kim, 2016). Finally, loyalty programs built into the VTO experience or granting exclusive access to new product launches can increase engagement and motivate conversion of utility-focused beauty products consumers.

6 Study Evaluation, and Suggestions for Further Research

While the research has shed light on the impact of VTO tools for beauty products on consumer behavior it is crucial to note the limitations on the generalizability of the findings imposed by sample size and its convenience nature. Although our sample is broader and larger than other studies related to VTO tools it is biased towards Gen Y and Z with the number of Gen X limited to just 5% of the sample. Utilizing panel data could have generated a more diverse sample. Cultural variables were not included in the study and should be examined in the future. Finally, only one VTO was used and, moreover, it was one belonging to a luxury brand. Both these conditions could have potentially biased the results. To control for these, multiple brands should be used as the quasi-experimental design prompts.

Nevertheless, this is one of the earliest studies on VTO characteristics and their effects on consumer behavior. In this sense, this chapter contributes to setting a solid framework for further exploration. Future studies can build upon it by exploring the applicability of other established theoretical frameworks.

Finally, to best of our knowledge this is the first research to consider socio-demographic and behavioral segmentation criteria. In this lies its originality. This first outline of the basic consumer segments of the online beauty retailing can, of course, be expanded and refined by considering a broader range of socio-demographics as well as personality traits and usage patterns. These could better guide the industry in its attempt to improve consumers' online shopping experience.

VTOs are increasingly becoming a subject of high interest in businesses aiming to advance their perceived experiential value and facilitate online sales. For beauty brands, embedding VTOs into their e-commerce platforms and integrating them into their marketing strategies as a key interactive communication tool might be the avenue to their future growth and profitability. The more hedonic and utilitarian consumer needs and tastes are accommodated the more consumers engage with brands and allow them to become parts of their social and personal lives.

Despite the fact that the fragrance sector of the beauty industry is its smallest one, representing an average of 17% of its total 2023 global value (but 27% in Europe), it is the fastest growing one at an annual rate of 14% and one expected to see accelerated growth in the 2023–2028 forecast period [2]. Maybe soon we will have a perfume VTO or we will see combined clothes, accessories, hairstyles and colors, and makeup VTOs. Until then, VTOs will be the spearhead of competitive advantage in the highly competitive makeup segment.

References

1. Berg, A., Hudson, S., Weaver, K.K., Pacchia, M.L., Amed, I.: The beauty market in 2023: a special State of Fashion report. McKinsey & Company (2023). https://www.mckinsey.com/industries/retail/our-insights/the-beauty-market-in-2023-a-special-state-of-fashion-report/
2. Weaver, K., et al.: The beauty boom and beyond: can the industry maintain its growth? Our Insights (2024). https://www.mckinsey.com/industries/consumer-packaged-goods/our-insights/the-beauty-boom-and-beyond-can-the-industry-maintain-its-growth
3. Doctor's Desk. Neuropeptides: The Future of Skincare Science in Clinikally Gurugram, India (2024). https://www.clinikally.com/blogs/news/neuropeptides-the-future-of-skincare-science
4. Best, W., Brown, M., Clark, T.: A year of known unknowns: the shifting policy landscape will define growth in 2025. U.S. Perspectives (2024). https://usa.visa.com/partner-with-us/visa-consulting-analytics/economic-insights/a-year-of-known-unknowns-the-shifting-policy-landscape-will-define-growth-in-2025.html
5. Eurostat. Living conditions in Europe - income distribution and income inequality (2024). https://ec.europa.eu/eurostat/statistics-explained/index.php?title=Living_conditions_in_Europe_income_distribution_and_income_inequality
6. Recalde, D., Jai, T.C., Jones, R.P.: I can find the right product with AR! The mediation effects of shopper engagement on intent to purchase beauty products. J. Retail. Consum. Serv. **78,** 103764 (2024). https://doi.org/10.1016/j.jretconser.2024.103764

7. Deshmukh, S., Sharma, A.F.: Analyzing consumer buying behavior in the context of search online, purchase offline in Indian council of social science research (ICSSR) National Conference on Social Entrepreneurship: Strengthening the Social Entrepreneurial Ecosystem through Innovation and Management Strategies (ed Thane Managing Committee Vidya Prasarak Mandal), pp. 63–66. BRIMS - Vidya Prasarak Mandal's Dr. V. N. Bedekar Institute of Management Studies, Thane, India (2024)
8. Liao, S.-H., Widowati, R., My Linh, P.N.: Delivery service and omni-channel online and offline in retailing during the Covid-19 pandemic. Sage Open **14**(4), 21582440241305047 (2024). https://doi.org/10.1177/21582440241305047
9. Watson, A., Alexander, B., Salavati, L.: The impact of experiential augmented reality applications on fashion purchase intention. Int. J. Retail Distrib. Manage. **48**(5) 433–451 (2018). https://doi.org/10.1108/ijrdm-06-2017-0117
10. Beck, M., Crié, D.: I virtually try it … I want it ! Virtual fitting room: a tool to increase on-line and off-line exploratory behavior, patronage and purchase intentions. J. Retail. Consum. Serv.s **40**, 279–286 (2018). https://doi.org/10.1016/j.jretconser.2016.08.006
11. Gabriel, A., Ajriya, A.D., Fahmi, C.Z.N., Handayani, P.W.: The influence of augmented reality on E-commerce: a case study on fashion and beauty products. Cogent Bus. Manage. **10**(2), 2208716 (2023). https://doi.org:https://doi.org/10.1080/23311975.2023.2208716
12. Whang, J.B., Song, J.H., Choi, B., Lee, J.-H.: The effect of Augmented Reality on purchase intention of beauty products: The roles of consumers' control. J. Bus. Res. **133**, 275–284 (2021). https://doi.org:https://doi.org/10.1016/j.jbusres.2021.04.057
13. Wang, Y., Ko, E., Wang, H.: Augmented reality (AR) app use in the beauty product industry and consumer purchase intention. Asia Pac. J. Mark. Logist. **34**(1), 110–131 (2022)
14. Trivedi, J., Kasilingam, D., Arora, P., Soni, S.: The effect of augmented reality in mobile applications on consumers' online impulse purchase intention: The mediating role of perceived value. J. Consum. Behav. **21**(4), 896–908 (2022). https://doi.org/10.1002/cb.2047
15. Hudson, S., Kim, A., Moulton, J.: What beauty players can teach the consumer sector about digital disruption in Our Insights. McKinsey & Company,Chicago (2018). https://www.mckinsey.com/industries/consumer-packaged-goods/our-insights/what-beauty-players-can-teach-the-consumer-sector-about-digital-disruption
16. Opute, A.P., Kalu, K.I., Chukwuma-Nwuba, E.O., Ojra, J., Iwu, C.G.: Technology and marketing: understanding the interface and Post-COVID-19 implications. In: Gbadamosi, A. (ed.) Critical Perspectives on Diversity, Equity, and Inclusion in Marketing, pp. 61–74. IGI Global (2022)
17. Kim, J., Forsythe, S.: Adoption of virtual try-on technology for online apparel shopping. J. Interact. Mark. **22**(2) 45–59 (2008). https://doi.org/10.1002/dir.20113
18. Rathee, V., Kumari, S.: Impact of virtual try-on technology on customer's mental imagery during COVID-19. IIMS J. Manage. Sci. **13**(1), 27–42 (2022)
19. Voicu, M.-C., Sîrghi, N., Toth, D.M.-M.: Consumers' experience and satisfaction using augmented reality apps in e-shopping: new empirical evidence. Appl. Sci. **13**(17), 9596 (2023). https://doi.org/10.3390/app13179596
20. Javornik, A.: Augmented reality: research agenda for studying the impact of its media characteristics on consumer behaviour. J. Retail. Consum. Serv. **30**, 252–261 (2016). https://doi.org/10.1016/j.jretconser.2016.02.004
21. Javornik, A.: 'It's an illusion, but it looks real!' Consumer affective, cognitive and behavioural responses to augmented reality applications. J. Mark. Manage. **32**(9–10), 987–1011 (2016). https://doi.org/10.1080/0267257X.2016.1174726
22. Lal, K.: We tested 7 virtual beauty try-on platforms to see if any of them are actually legit: VTO's foothold on the beauty industry is only getting stronger—but which platforms really deliver? in Fashionista (2023). https://fashionista.com/2023/11/virtual-beauty-makeup-try-on-online-apps

23. Plotkina, D., Saurel, H.: Me or just like me? The role of virtual try-on and physical appearance in apparel M-retailing. J. Retail. Consum. Serv. **51**, 362–377 (2019). https://doi.org/10.1016/j.jretconser.2019.07.002
24. Kavilanz, P.: No testers, no problem: ulta and Sephora have a new take on 'try before you buy'. CNN Business (2020)
25. L'Oréal. L'Oréal's modiface brings AI-powered virtual makeup try-on to amazon in articles / science and technology. L'Oréal Group. https://www.loreal.com/en/articles/science-and-technology/l-oreal-modiface-brings-ai-powered-virtual-makeup-try-ons-to-amazon/
26. Sekri, K., Bouzaabia, O., Rzem, H., Juárez-Varón, D.: Effects of virtual try-on technology as an innovative e-commerce tool on consumers' online purchase intentions. Eur. J. Innov. Manage. (2024)
27. Arthur, R.: Charlotte Tilbury's new virtual 'Magic Mirror' serves as active make-up Sellingtool. Forbes (2016). https://www.forbes.com/sites/rachelarthur/2016/10/21/charlotte-tilburys-new-virtual-magic-mirror-serves-as-active-make-up-selling-tool/
28. Tilbury, C.: Try-on makeup magic with Charlotte's magic mirror makeup online try on! (2020). https://www.charlottetilbury.com/us/secrets/magic-mirror-makeup-online-try-on
29. Market Reports. Virtual makeup try-on market, global outlook and forecast 2023–2032. New York, NY (2023)
30. Bonetti, F., Warnaby, G., Quinn, L.: Augmented reality and virtual reality in physical and online retailing: a review, synthesis and research agenda. In: Jung, T., Claudia tom Dieck, M. (eds.) Augmented Reality and Virtual Reality: Empowering Human, Place and Business, pp. 119–132. Springer (2018)
31. Nawaz, M.Z., Guzmán, F., Nawaz, S.: Technology-enabled engagement process of brand virtual-try-on services. J. Prod. Brand Manage. **34**(1), 44–60 (2025). https://doi.org/10.1108/JPBM-12-2023-4859
32. Javornik, A., Marder, B., Pizzetti, M., Warlop, L.: Augmented self - the effects of virtual face augmentation on consumers' self-concept. J. Bus. Res. **130**, 170–187 (2021). https://doi.org/10.1016/j.jbusres.2021.03.026
33. Yang, X.: Augmented reality in experiential marketing: the effects on consumer utilitarian and hedonic perceptions and behavioural responses. In: Lee, Z.W.Y., Chan, T.K.H., C.M.K. (eds.) Information Technology in Organisations and Societies: Multidisciplinary Perspectives from AI to Technostress, pp.147–174. Emerald Publishing Limited (2021)
34. Baek, T.H., Yoo, C.Y., Yoon, S.: Augment yourself through virtual mirror: the impact of self-viewing and narcissism on consumer responses. Int. J. Advert. **37**(3), 421–439 (2018). https://doi.org/10.1080/02650487.2016.1244887
35. Kumar, H.: Augmented reality in online retailing: a systematic review and research agenda. Int. J. Retail Distrib. Manage. **50**(4), 537–559 (2022). https://doi.org/10.1108/IJRDM-06-2021-0287
36. Al Hilal, N.S.H.: The impact of the use of augmented reality on online purchasing behavior sustainability: the Saudi consumer as a mode. Sustainability **15**(6), 5448 (2023). https://doi.org/10.3390/su15065448
37. Efstathiou, E., Drossos, D.: Augmented reality marketing: factors that affect the intention to use a virtual try-on mobile application, pp. 32–46. Springer (2024)
38. Lee, H., Xu, Y., Porterfield, A.: Antecedents and moderators of consumer adoption toward AR-enhanced virtual try-on technology: a stimulus-organism-response approach. Int. J. Consum. Stud. **46**(4), 1319–1338 (2022). https://doi.org/10.1111/ijcs.12760
39. Zhang, T., Wang, W.Y.C., Cao, L., Wang, Y.: The role of virtual try-on technology in online purchase decision from consumers' aspect. Internet Res. **29**(3), 529–551 (2019). https://doi.org/10.1108/intr-12-2017-0540
40. Kokku, R.: Omni and multi-channel: relationship with utilitarian/hedonic benefits, shopping value and channel patronage. Int. Rev. Manag. Mark. **11**(2), 11–22 (2021)

41. Yim, M.Y.-C., Chu, S.-C., Sauer, P.L.: Is augmented reality technology an effective tool for E-commerce? An interactivity and vividness perspective. J. Interact. Mark. **39**, 89–103 (2017). https://doi.org/10.1016/j.intmar.2017.04.001
42. Ajzen, I.: Attitude structure and behavior. In: Attitude Structure and Function. The third Ohio State University Volume on Attitudes and Persuasion, pp. 241–274. Lawrence Erlbaum Associates, Inc. (1989)
43. Hilken, T., Keeling, D.I., de Ruyter, K., Mahr, D., Chylinski, M.: Seeing eye to eye: social augmented reality and shared decision making in the marketplace. JAMS **48**(2), 143–164 (2020). https://doi.org/10.1007/s11747-019-00688-0
44. Zarkada, A.K., Panigyrakis, G.G., Tzoumaka, E.: Hosting a successful metamodern party: mixed methods management research on the Web 2.0+. In: Moutinho, L., Sokele, M. (eds.) Innovative Research Methodologies in Management Vol. I: Philosophy, Measurement and Modelling, Ch. 1, pp. 1–29. Palgrave Macmillan (2018)
45. Zarkada, A.K.: Listening to a unicorn's pink narrative: scheduling of brand posts on instagram for maximum consumer brand engagement. In: 19th International Conference on Research in Advertising (ICORIA) Virtual Conference. European Advertising Academy (2021)
46. Joshi, A., Kale, S., Chandel, S., Pal, D.K.: Likert scale: explored and explained. Curr. J. Appl. Sci. Technol. **7**(4), 396–403 (2015). https://doi.org/10.9734/BJAST/2015/14975
47. Nunan, D., Malhotra, N.K., Birks, D.F., Malhotra, N.: Marketing Research: Applied Insight. Pearson (2020)
48. WebAIM. WAVE Web Accessibility Evaluation Tool (2024). https://wave.webaim.org/
49. Sam-Epelle, I., Olayinka, O., Jones, P.: The evolution of enterprise gamification in the digital era and the role of value-based models. Sustainability **14**(15), 9251 (2022). https://doi.org/10.3390/su14159251
50. Kohler, E., Mogaji, E., Erkan, İ: Save the trip to the store: sustainable shopping, electronic word of mouth on instagram and the impact on cosmetic purchase intentions. Sustainability **15**(10), 8036 (2023). https://doi.org/10.3390/su15108036
51. Hilken, T., de Ruyter, K., Chylinski, M., Mahr, D., Keeling, D.I.: Augmenting the eye of the beholder: exploring the strategic potential of augmented reality to enhance online service experiences. JAMS **45**(6), 884–905 (2017). https://doi.org/10.1007/s11747-017-0541-x
52. Kuehnl, C., Jozic, D., Homburg, C.: Effective customer journey design: consumers' conception, measurement, and consequences. JAMS **47**(3), 551–568 (2019). https://doi.org/10.1007/s11747-018-00625-7
53. Kowalczuk, P., Siepmann, C., Adler, J.: Cognitive, affective, and behavioral consumer responses to augmented reality in e-commerce: a comparative study. J. Bus. Res. **124**, 357–373 (2021). https://doi.org/10.1016/j.jbusres.2020.10.050

AI, Automation, and Digital Transformation in Industry- Specific Applications

Information Systems and Marketing of Tour Operators: Impacts, User Perspectives, Challenges, and Future Directions

Anna Krousaraki[1(✉)] and Ourania Vitouladiti[2]

[1] Hellenic Open University, Patras, Greece
krousarakianna1@gmail.com
[2] University of West Attica, Egaleo, Greece
ranivito@uniwa.gr

Abstract. Tour Operators have significantly benefited from the adoption of advanced information systems and technological innovations. These systems have aided in improving reservation management, creating tourism packages, and developing marketing strategies, etc. This study aims, through a literature review, to document the characteristics and ongoing developments of information systems in tourism. Additionally, it seeks to examine the impacts of advanced information systems on various marketing functions. However, these elements represent only one side of the equation. The other side concerns the users of these systems, who are responsible for applying their capabilities and consequently evaluating their features, advantages, and disadvantages within the marketing field. This aspect of the topic, based on the literature review, has not been analyzed in a way that adds knowledge to this scientific field or lays the foundation for potential improvements and further considerations. This specific approach constitutes the focus of the research part of this study. The findings of the study confirm the points mentioned in the literature review and reveal the evaluation of the criteria for acquiring an information system, its advantages, and weaknesses in the marketing field, based on the importance attributed by the system users themselves. The respondents perceive the existence of obstacles in utilizing information systems, particularly in issues of personalizing experiences and contributing to the success of tourism activities. The employees highlight the need for continuous upgrades and development, as well as the importance of the human factor.

Keywords: Information Systems · Tour Operators · Marketing · Challenges in Tourism · Digital Transformation in Tourism

1 Introduction

The tourism sector, as a fundamental pillar of economic activity and cultural exchange, has undergone drastic changes due to information systems. This study delves into the core of this transformation, focusing on Tour Operators (T.O.), who emerge as the architects of travel experiences. The motivation behind this research stems from the recognition

A. K. Zarkada (Ed.): GMA-GAMMA 2025, SPBE, pp. 125–153, 2025.
https://doi.org/10.1007/978-3-031-96930-0_8

that, in an era dominated by digital technology, Tour Operators face countless challenges as well as opportunities in their quest for continuous development and sustainability. This makes the research both highly relevant and intriguing.

This study explores, through a literature review, the characteristics and properties of information systems and the multifaceted challenges faced by tourism enterprises. It emphasizes the need for effective management systems, adaptability to traveler demands, and skilled handling of changes, introducing considerations for their more efficient management. Additionally, the analysis of the integration of information systems into marketing field reveals how these systems shape business processes, offering a comprehensive view of the mutual relationship between technology and the fundamental elements of tourism strategy.

At the research level, an effort is made to initially record the user's perspective on a series of criteria for the selection of information systems by a tour operating company. Subsequently, within the framework of marketing functions, an attempt is made to evaluate the user's assessment of the importance of the systems' advantages, their weaknesses, and the challenges arising from their use today, where the tourist-consumers are more demanding regarding the products and services they desire and, consequently, the experiences they seek.

Therefore, the objectives of this study are:

1. Regarding the literature review:
   - To record the characteristics and ongoing developments of information systems in the tourism sector.
   - To provoke considerations regarding the management of these developments by Tour Operators.
   - To examine the impacts of advanced information systems on various functions of Marketing.
2. Regarding the primary research:
   - To record and analyze the importance of the criteria for selecting an information system from the user's perspective.
   - To record and analyze the importance of the benefits and problems (within the framework of marketing functions) arising from the use of an information system from the user's perspective.
   - To record and analyze, through cross tabulation of specific variables, the users'/employees' perspectives (by department within a tour operating company) on the importance of benefits for more effective Marketing strategies through the use of an information system.
   - To record and analyze, through cross tabulation of specific variables, the users'/employees' perspectives (by department within a tour operating company) on the importance of benefits for the promotion of the company through the use of an information system.
   - To record and analyze, through cross tabulation of specific variables, the users'/employees' perspectives (by department within a tour operating company) on the importance of problems arising from the difficulty of adapting products and services to the new needs of travelers.

The research approach aims to create additional literature that incorporates challenges, considerations, benefits, and problems, offering scholars further perspectives in a rapidly evolving field. At the same time, it seeks to provide the professional sector with information to improve their operations and resolve issues arising from the use of information systems.

2 Definition and Classification of Information Systems

Tour Operators must continuously adapt to changes, preferences, and the needs of travelers to ensure their survival [1]. A critical aspect of this adaptation is the integration of information systems. Information systems play a significant role in the tourism sector, providing numerous advantages in tourism transactions. Information technologies have improved the speed, efficiency, and global reach of the travel industry while also contributing to cost reduction. The use of information systems enables Tour Operators to offer a broader range of products through interconnected interfaces, known as 'Network Economies.' This real-time connectivity benefits businesses by linking them with suppliers, operational processes, and customers [2].

The continuous integration of information systems in tourism enhances speed, reduces costs, and improves reliability. Tour Operators who utilize these systems can effectively coordinate their operations, facilitating tasks such as the creation, booking, and pricing of various tour components. The technology-based approach also leads to more efficient and error-free transactions through the adoption of services that rely on the use of additional software [3].

2.1 Information Systems

What are information systems? Information systems constitute a complex combination of technology and human activity. One definition of information systems states that they represent a set of interacting components that collect (or retrieve), process, store, and distribute information to support decision-making and control in an organization. Generally, it could be argued that their primary purpose is to assist individuals working with data to improve a collective activity [4]. Olivé identifies three main functions of information systems: the memory function, the informative function, and the active function. Consequently, an information system records representations of the general state of the system, provides information about its status, and supports the execution of actions that change its state [5].

2.2 Information Systems for Tour Operators

Tourism information systems constitute an essential pillar for Tour Operators (T.O.). They effectively undertake the core functions of collecting, processing, storing, and disseminating information related to tourism, thereby providing significant support to the activities of T.O. Through these systems, T.O. can efficiently manage information about tourism packages, personalize their offerings, and provide more specialized services to travelers. In this way, tourism information systems are an integral part of the modern tourism sector and contribute to the promotion of innovation and competitiveness of T.O. [4].

Tourism information systems include various subsystems and applications used by Tour Operators for managing and providing services. Specifically, they include:

1. Computer Reservation Systems and Global Distribution Systems (CRS and GDS)
2. Hotel Management Systems
3. Accounting and Financial Management Systems

2.3 Reasons Why Tour Operators Utilize Information Systems

Computer Reservation Systems and Global Distribution Systems (CRS and GDS) are platforms that enable Tour Operators to manage bookings in real-time. These systems provide access to a wide range of travel services, such as flights, hotels, car rentals, and other tourism activities. By using these systems, Tour Operators can offer their clients immediate and accurate information, enhancing the customer experience and increasing their satisfaction.

Hotel Management Systems are platforms that assist in managing the daily operations of hotels. These systems include functions such as reservation management, room division management, customer service, and inventory management. By utilizing these systems, hotels can improve operational efficiency, reduce costs, and provide better services to their customers. The integration of the central reservation system with Global Distribution Systems (GDS) fosters a harmonious relationship between hotels and Tour Operators (T.O.). These systems serve as the link connecting hotels with T.O., enhancing a collaborative "ecosystem".

Accounting and Financial Management Systems enable the effective management of financial transactions and the monitoring of financial data. These systems assist Tour Operators in managing their revenues and expenses, tracking payments and receipts, and ensuring the financial stability of their business.

These systems are crucial for Tour Operators (T.O.) as they enhance efficiency, reduce costs, and increase the reliability of their services. By integrating these systems, T.O. can offer better services to their clients, respond more quickly to their needs, and maintain their competitiveness in the market.

Global Distribution Systems

Among the information systems in tourism, with a focus on those relevant to Tour Operators, Global Distribution Systems (GDS) play a fundamental and specialized role due to the breadth of the services they offer. Therefore, it is essential to make a particular reference to Global Distribution Systems (GDS) and the reasons why Tour Operators utilize them.

Tour Operators utilize Global Distribution Systems (GDS) for various reasons, leveraging a central system that optimizes different aspects of their business activities. Some of these reasons include:

1. Extensive and immediate access to information: GDS (Global Distribution Systems) establish connections with a wide network of travel service providers, including airlines, hotels, car rental agencies, and other travel-related services. This extensive

access enables Tour Operators to present a comprehensive range of travel products and services to their clients, offering a unified solution for various travel needs, such as creating travel packages. For example, there is immediate access to information regarding airlines, available flight information from all companies, flight availability, and more [6].

2. Improved booking processes: GDS (Global Distribution Systems) facilitate quick and real-time booking procedures by providing Tour Operators (T.O.) with immediate access to updated information regarding availability and prices from travel service providers. The streamlined booking process allows T.O. to efficiently confirm and manage reservations for their clients. Continuing with the example of airlines, there is immediate access to information about seat availability, prices, and the possibility of online purchase [6].
3. Global reach and connectivity: GDS (Global Distribution Systems), with their worldwide presence, connect Tour Operators (T.O.) to an extensive network of international travel service providers. This global reach enables T.O. to curate and offer travel packages that span multiple destinations, catering to the diverse preferences and interests of their clients [7].
4. Customer database creation: The use of GDS (Global Distribution Systems) also facilitates the creation and management of a comprehensive customer database. Through real-time access to information and improved booking processes, Tour Operators (T.O.) can collect valuable data on customer preferences, booking history, and travel behaviors. By leveraging extensive access to information regarding availability and connectivity offered by GDS, T.O. can tailor their services to meet the individual needs of customers, enhancing their satisfaction and loyalty [6].
5. Marketing and promotion opportunities: GDS (Global Distribution Systems) serve as effective marketing platforms for travel service providers, and Tour Operators can leverage the promotional opportunities within GDS to showcase their travel packages to a broader audience [7].

In the modern digital landscape, Tour Operators (T.O.) without a strong online presence risk falling behind in developments. Contemporary travelers prefer the convenience of browsing and booking travel services online, where they are provided with the ability to make last-minute reservations in real-time. The evolution of digital marketing has played a crucial role in simplifying communication and overcoming barriers in the tourism and hospitality sectors. Communication technologies and modern digital marketing methods have successfully addressed these issues, enhancing interactions between tourists and the hospitality industry. The use of GDS (Global Distribution Systems) enables T.O. to extend their reach, increase visibility, streamline distribution, and gain a competitive advantage in the tourism sector [8].

2.4 Considerations Regarding the Implementation of Information Systems in Tour Operators

Tour Operators face the challenge of selecting systems that not only meet their operational needs but also adapt to the ever-changing market demands. The adoption of new technology can raise concerns regarding cost, staff training, and ongoing support.

Additionally, the need for systems that offer high usability and continuous upgrades is critical to ensuring efficiency and user satisfaction. These concerns guide businesses in seeking solutions that will enhance their competitiveness and ensure their sustainability in a dynamic environment.

3 The Impact of Information Systems on the Marketing of Tour Operators

Marketing plays an increasingly significant role in the tourism sector. For industry professionals, marketing means understanding customers and developing a system of increasingly personalized service delivery that offers desired products at reasonable prices and high quality, while ensuring the profitability of the business. The rapid evolution of marketing has given customers a powerful voice, allowing them to express their opinions and influence the market. Additionally, Millennials, who replace postwar generations as the most important segment of the tourism market, have different desires and expectations. Technology in the field of marketing for large tourism enterprises has created a competitive environment where understanding and adapting to customer needs is critical for success. Overall, marketing is an integral part of a Tour Operator's strategy for achieving sustainable growth and success in the tourism sector [9].

3.1 The Benefits of Using Advanced Information Systems by Tour Operators in the Field of Marketing

Tour Operators (T.O.) are significantly experiencing the benefits of digitizing tourism activities. Traditional practices have undergone transformation due to the changes implemented by T.O. With the adoption of "new" and advanced information systems, T.O. are experiencing a substantial transformation in the way they design and execute their activities. This shift is evidence of effective change management strategies, ensuring a smooth transition to modernized practices [10].

The use of advanced information systems by Tour Operators offers multifaceted advantages, particularly in enhancing customer experience and boosting business visibility. The implementation of these advanced systems allows operators to reduce manual workload, providing an upgraded experience for both customers and the business itself. This not only enhances efficiency but also saves valuable time.

Specifically, according to various authors the key advantages that arise from the use of information systems are [12–19]:

1. Enhanced customer experience: Advanced information systems enable Tour Operators (T.O.) to tailor their products based on customer preferences, thereby enhancing overall satisfaction and loyalty. Through personalized interactions facilitated by these systems, customers actively co-create their consumption experiences, aligning with modern trends such as experiential marketing. This personalization fosters a deeper emotional connection between customers and T.O., leading to long-term commitment and positive word-of-mouth promotion. By leveraging data analysis and customer insights derived from advanced information systems, T.O. can anticipate and respond

to evolving customer needs in real time. This approach not only enhances the relevance and appeal of tourism experiences but also strengthens the sense of exclusivity and differentiation, positioning T.O. as "preferred" in an increasingly competitive market.

2. Enhanced insights for Tour Operators (T.O.): Advanced information systems provide T.O. a wide range of information, from customer preferences to market trends and competitor activities. This increased visibility empowers T.O. to make data-driven decisions and strategic investments that optimize resource allocation and maximize return on investment. By leveraging these analytical insights, T.O. can anticipate market changes and emerging trends, allowing them to adapt accordingly.
3. Improved operations: Advanced information systems streamline the internal processes and workflow of Tour Operators (T.O.), enhancing the efficient use of resources. Automated booking systems, integrated marketing platforms, and real-time communication channels enable coordination among departments and stakeholders, minimizing delays and errors. Additionally, by centralizing data management and leveraging cloud-based technologies, T.O. can access information anytime, anywhere, facilitating remote collaboration and decision-making. This flexibility is particularly valuable in dynamic environments where quick adaptation is essential for survival and growth.
4. Targeted Marketing and Personalization: Advanced information systems enable Tour Operators (T.O.) to effectively segment their customer database and offer targeted marketing campaigns tailored to specific demographics, preferences, and behaviors. By utilizing big data analytics algorithms, T.O. can identify lucrative market segments and create personalized experiences that resonate with their target audience. Additionally, by integrating customer relationship management (CRM) systems with social media platforms and digital channels, T.O. can engage customers in meaningful dialogues and enhance brand reputation. This interactive approach not only boosts customer satisfaction and loyalty but also increases brand visibility and reputation in the digital marketplace.

Example of the Use of Advanced Information Systems by Tour Operators in the Field of Marketing

Ghana has positioned tourism as a part of its governmental policy. Two Ghanaian Tour Operators (T.O.), Apstar Tours Limited and Ashanti Africa Tours, have already taken steps in adopting modern information systems and utilizing innovative marketing strategies to enhance the country's status as a sought-after destination. By using advanced information systems, they have streamlined booking processes, improved communication channels, and offered personalized experiences to visitors. Their digital marketing efforts, including engagement on social media, have boosted Ghana's tourism brand visibility, attracting a diverse audience of travelers. By integrating these approaches, Ghana's tourism sector has created growth opportunities, achieved repeat visitors, and is considered a top destination in West Africa [20].

In conclusion, the adoption of advanced information systems by Tour Operators (T.O.) offers a multitude of advantages, ranging from enhanced customer experience and improved business visibility to optimized operations and targeted marketing. By leveraging the power of data-driven insights, T.O. can confidently be regarded as industry leaders, capable of delivering exceptional value to their clients while simultaneously driving sustainable growth and competitiveness in the global tourism landscape [11].

3.2 Capabilities and Prospects in the Field of Marketing from the Implementation of Information Systems in Tour Operating

Customer Approach and Service Quality in the Digital Tourism Sector
Smart tourism emphasizes the provision of high-quality services through the application of Information and Communication Technologies (ICT), ensuring seamless interactions and enriching customer experiences. By leveraging digital technologies, Tour Operators can optimize resource utilization and promote sustainable development [21]. Smart tourism transforms the traveler's experience at all stages of their journey, from pre-trip planning to post-trip evaluation. Digital technologies enable travelers to explore numerous options, access real-time information, and interact with destinations and fellow travelers through social media and mobile applications. Augmented and virtual reality experiences provide effective previews of destinations, while continuous connectivity allows for seamless interaction and content sharing [21].

Creation of Innovative Products Using Information Systems in Tour Operating
Information systems play a central role in promoting innovation within organizations. By recognizing information as a strategic asset, Tour Operators leverage acquired knowledge to gain a competitive advantage. Through the integration of information systems, Tour Operators motivate employees to unleash their creative potential, thereby leading to continuous excellence. These systems are essential for facilitating decision-making and planning, which are vital for innovation. The synergy between information systems and innovation streamlines processes, enhances collaboration and accelerates the development of new ideas. Essentially, information systems serve as catalysts for organizational innovation, driving growth and success in dynamic markets [22].

The creation of innovative tourism products and services should be undertaken by Tour Operators through the integration of modern technologies in the development and management processes of tourism products. Tour Operators can benefit from information systems for effective monitoring of tourism trends, customer preferences, and performance metrics. These systems facilitate market data analysis, customer management, destination management, marketing campaign implementation and more [23].

Innovative Ideas for Tourism Products Using AI-Powered Technology
The integration of AI technology for the creation of innovative products in tourism has revolutionized the sector by enhancing personalized experiences, optimizing operations, and expanding the reach of destinations. From chatbots assisting travelers to the use of artificial intelligence (AI), technology continues to reshape and enrich the tourism landscape, offering unparalleled opportunities for growth and progress. Some examples according to Wilson [24] include:

1. Luxury Tourism: Creating personalized experiences and services for travelers seeking "luxurious" experiences. AI-powered personal assistants and virtual hotel advisors enhance the stay experience.

2. Smart Destination Planning: Utilizing data and predictive analytics to enable tourists to plan their itineraries. Real-time data analysis on weather, traffic, tourist products, and destinations facilitates informed decisions and avoids overcrowded areas.
3. Adventure Tourism: Unique adventure experiences that highlight natural beauty, such as hiking, camping, climbing, etc. Adventure is enhanced with digital technology, including wearable trackers, drone footage, and more.

Digital Platforms and the Creation of Distribution Channels
In recent years, digital platforms have emerged as an even more essential means for providing services in the tourism sector. These platforms offer a wide range of services, including hotel reservations, flights, travel packages, attraction tickets, cruises, and car rentals [25].

Hotel reservations have particularly thrived through the internet, establishing it as the primary channel for potential visitors to access information and plan their trips without direct contact with Tour Operators (T.O.). One of the most significant outcomes of using digital platforms as distribution channels is that they can help T.O.s acquire a larger and more diverse customer base, both locally and globally. They can tap into new markets, segments, and specialties, and tailor their offerings to different needs, preferences, and customer behaviors. Digital platforms help increase the recognition of T.O.s and create a distinct identity. Additionally, the use of digital platforms as distribution channels aids in data generation, the analysis of which can assist in creating innovative products [26].

3.3 Challenges in Digitalization within the Field of Marketing

In the field of smart tourism, Tour Operators (T.O.) face several challenges that require careful navigation and strategic planning. According to various authors, these challenges include [27–33]:

1. The shift towards digitalization requires Tour Operators (T.O.) to reconsider traditional models and adopt innovative strategies to effectively meet the needs of digital tourists. Leveraging data-driven insights and technology-based marketing strategies presents both opportunities and challenges for businesses. The emergence of virtual reality technologies necessitates innovative approaches to engage modern travelers.
2. The challenge of infrastructure and network is significant, as T.O.s must ensure ubiquitous access to internet technology at various points of interest to meet travelers' demands. The availability of effective Wi-Fi networks is crucial for providing real-time access to information and enhancing visitor experiences, especially since many of the information travelers seek requires access during their actual journey rather than during the planning phase.
3. Integrating marketing processes such as SoCoMo (Social, Contextual, and Mobile) Marketing poses a challenge for T.O.s aiming to effectively attract travelers. Implementing marketing strategies that utilize social media, location-based services, and mobile capabilities require careful orchestration to create personalized and effective interactions.
4. T.O.s must prioritize security and sustainability in their technological efforts to ensure their reputation and financial security.

By confronting these challenges head-on, Tour Operators (T.O.) can enhance their competitiveness in the global tourism market and promote sustainable development within the smart tourism ecosystem [21].

An example of how travelers are affected by recent innovations is the "digital travel friend". This is an application available on customers' smartphones that can manage all the details of their trip during their journey [21].

3.4 Considerations for the Adaptation of Tourism Enterprises in the Era of Advanced Technologies and in Relation to Marketing Activities

The continuous evolution of the tourism sector and emerging technological advancements present significant considerations for Tour Operators. How can they effectively leverage advanced information systems to enhance customer experience? What are the best practices for adapting marketing strategies and creating products that address the changing needs of travelers? How can they efficiently manage digital distribution channels and online platforms while maintaining their competitiveness and meeting buyer demands? How can they develop successful strategies that allow them to adapt to the constant changes of the digital environment and capitalize on the advantages offered by the digital age for the successful growth of their businesses?

The use of information systems in Tour Operators' businesses is a critical issue that requires careful analysis and evaluation. Information systems can offer multiple benefits, such as more effective marketing strategies, enhanced business visibility, customer engagement, increased sales, improved customer satisfaction, creation of new products, increased visibility in digital distribution channels, synergy for creating tourism packages, and synergy for managing bookings. However, it is important to examine to what extent these potential benefits are confirmed in practice by the users themselves.

The use of information systems in Tour Operators' businesses presents significant challenges that require careful analysis and evaluation. Despite the multiple benefits they offer, such as the improvement of marketing strategies and increased customer satisfaction, there are also serious issues that need to be addressed. These include the difficulty in adapting to changes in the digital environment, monitoring changes in traveler preferences, providing personalized experiences, adapting products and services to new needs, adjusting marketing strategies, and the ineffective management of distribution channels.

These considerations highlight the need for continuous exploration of solutions and innovations that will enable Tour Operators to remain competitive and progressive in a dynamically changing sector.

3.5 Research Questions

The literature review reveals the rapid evolution of the characteristics of information Systems in tour operating and the necessity of evaluating the respective criteria for a business to select a system. At the same time, it highlights the existence of benefits, problems, and challenges arising from the use of information Systems.

Therefore, given their characteristics and the existence of both benefits and problems, a series of research questions arise regarding the evaluation of these characteristics, benefits, and problems. In this case, the ideal respondents for the previously referenced evaluation would be the users themselves. This approach lends originality and differentiation to research compared to similar studies that focus on listing characteristics and developments in the field.

Therefore, the research questions that arise are:

1. What is the degree of importance that users attribute to various criteria when selecting an information system by a Tour Operator?
2. What is the degree of importance that users attribute to the various benefits from the use of information systems in tour operators' businesses?
3. What is the degree of importance that users attribute to the various problems arising from the use of information systems in tour operators' businesses?
4. What is the degree of importance that users attribute to the benefits of using information systems for more effective marketing strategies, in relation to the department in which they work?
5. What is the degree of importance that users attribute to the benefits of using information systems for the promotion of businesses, in relation to the department in which they work?
6. What is the level of importance that users consider for the problems arising from the difficulty of adapting products and services to the new needs of travelers, due to the use of information systems, in relation to the department in which they work? The previously referenced questions also define the research objectives of the present study.

4 Research Methodology

4.1 Research Objectives

The objectives of this research are, from the user's perspective, to evaluate the importance of the criteria for selecting an information system, the importance of the benefits, and the problems associated with the use of these systems, with reference to marketing variables.

Furthermore, through the cross tabulation of variables, the study aims to record and analyze the perspectives of users/employees (by department within a tour operating company) regarding the importance of benefits for more effective marketing strategies and specifically for the promotion of the company through the use of an information system. Additionally, it aims to record and analyze, again through the cross tabulation of variables, the perspectives of users/employees (by department within a tour operating company) regarding the importance of problems arising from the difficulty of adapting products and services to the new needs of travelers.

4.2 Research Methodology and Research Tool

The research methodology was based on primary quantitative research, with the main research tool being the questionnaire. The questionnaire was anonymous to ensure the confidentiality of the participants, and the questions were closed-ended, using five-point scales to evaluate the variables. The creation, management, and distribution of the questionnaire were conducted through the Microsoft Forms platform, while the analysis of the primary data was performed using Microsoft Excel software.

4.3 Description of the Sample

The distribution of the questionnaire was conducted among employees of leading tour operators in 18 countries. The participating countries included Greece, Spain, Thailand, Mexico, Cyprus, Bulgaria, France, Morocco, Portugal, Turkey, Croatia, Italy, Tunisia, United Kingdom, Costa Rica, Dominican Republic, Germany, and Switzerland. The period was April 2024. The total number of questionnaires was 133. Through this process, data were collected from various tourism markets, thereby enhancing the validity of the study's results [1].

5 Results

5.1 Basic Information of the Research Participants

Table 1. Job Position in the company

	N	%
Operations Support	48	36.1%
Operations	31	23.3%
Finance, Accounting and HR	16	12.0%
Category Management	15	11.3%
Product Development and Planning	11	8.3%
Technology and IT	7	5.3%
Sales and Marketing	4	3.0%
Customer Service	1	0.8%
Total	**133**	**100.0%**

(Source: Krousaraki, 2024).

The analysis of the data from the 133 questionnaires regarding the "Job Position in the company", as depicted in Table 1, reveals that the Operations Support department has the majority of participants, with 48 respondents (36.1%). The Operations department follows with 31 respondents (23.3%). Additionally, 54 questionnaires make up the remaining 40.6% of the sample, covering a wide range of other departments. The fact that the Operations departments hold the majority of participants is expected, as they traditionally constitute the majority of employees in tour operators, which adds credibility to the present study.

The analysis of the data concerning the "Current Job Title", as depicted in Table 2, shows that the Manager - Lead position holds the highest number of participants, with 66 respondents (49.6%). The Executive – Agent position follows with 56 respondents (42.1%). It is important to note that the sample is well distributed, as it includes representatives from both higher management positions and lower hierarchical positions.

Table 2. Current Job Title

	N	%
Manager - Lead	66	49.6%
Executive – Agent	56	42.1%
Analyst	8	6.0%
Director - Head	3	2.3%
Total	**133**	**100.0%**

(Source: Krousaraki, 2024).

Table 3. Country

	N	%
Greece	46	34.6%
Spain	27	20.3%
Thailand	10	7.5%
Mexico	9	6.8%
Cyprus	6	4.5%
Bulgaria	5	3.8%
France	4	3.0%
Morocco	4	3.0%
Portugal	4	3.0%
Turkey	4	3.0%
Croatia	3	2.3%
Italy	3	2.3%
Tunisia	2	1.5%
UK	2	1.5%
Costa Rica	1	0.8%
Dominican Republic	1	0.8%
Germany	1	0.8%
Switzerland	1	0.8%
Total	**133**	**100.0%**

(Source: Krousaraki, 2024).

This enhances the reliability of the results and provides a more comprehensive view of the distribution of job positions.

The distribution of the questionnaire was conducted internationally with the aim of capturing data from professionals employed by tour operators worldwide. Of the 133 questionnaires, as recorded in Table 3, 46 (34.6%) were from employees in Greece, while 87 questionnaires (65.4%) were from employees in other countries. This indicates a sample with significant geographical diversity.

5.2 Demographic Characteristics of the Research Participants

Table 4. Age

	N	%
Up to 25	4	3.0%
25–34	22	16.5%
35–44	44	33.1%
45–54	40	30.1%
55 and above	23	17.3%
Total	133	100.0%

(Source: Krousaraki, 2024).

The data concerning "Age", as recorded in Table 4, indicate that the largest age group of employees in the Tour Operators sector who are active users of information systems is the 35–44 age group, representing 33.1%, closely followed by the 45–54 age group, representing 30.1%. This suggests that the employees in this sample have accumulated experience and are likely in more mature phases of their professional careers. This adds credibility to the findings.

Table 5. Gender

	N	%
Female	90	67.7%
Male	40	30.1%
Prefer not to say	3	2.3%
Total	**133**	**100.0%**

(Source: Krousaraki, 2024).

The conclusions drawn from the analysis of the data in Table 5 regarding "Gender" are as follows: Women (90 participants, representing 67.7%) constitute the majority of the workforce in the sample. This may reflect employment trends in the Tour Operators sector. Men represent the minority of the workforce in the sample (40 participants, representing 30.1%).

The data concerning the "Educational level", as recorded in Table 6, indicate that the majority of participants have a Bachelor's degree (50.4%), a significant percentage

Table 6. Educational level

	N	%
Bachelor's degree	67	50.4%
Master's degree	31	23.3%
High school diploma	18	13.5%
Vocational/Technical training	15	11.3%
Doctorate degree	2	1.5%
Total	**133**	**100.0%**

(Source: Krousaraki, 2024).

also have a Master's degree (23.3%), and a small percentage (1.5%) have doctoral studies. Overall, this suggests that the sample consists of individuals with a high level of education.

Table 7. Studies in Tourism

	N	%
No	77	57.9%
Yes	56	42.1%
Total	**133**	**100.0%**

(Source: Krousaraki, 2024).

The data in Table 7 indicate that 57.9% of the employees do not have a specialization in tourism, while 42.1% have studies related to tourism. The presence of employees with a specialization in tourism is significant, as these employees may have a deeper understanding of tourism products, markets, and customers, among other aspects.

Table 8. Years of experience in the tourism sector

	N	%
More than 10 years	87	65.4%
6–10 years	25	18.8%
1–5 years	20	15.0%
Less than 1 year	1	0.8%
Total	**133**	**100.0%**

(Source: Krousaraki, 2024).

The data in Table 8 concerning "Years of experience in tourism sector" indicate that 65.4% of the sample consists of employees with more than 10 years of experience. This highlights that the sample includes a significant number of employees with extensive work histories, thereby enhancing the validity of the results. The data reflects the views and perceptions of an experienced workforce, suggesting that the findings can be considered representative of the actual conditions and practices prevailing in the sector.

5.3 Tour Operators and Information Systems in Tourism

The **first** question of this section aims to assess the level of knowledge and familiarity of employees in Tour Operators with information systems. It is a closed-ended question that asks participants to rate their personal experience with information systems. This question refers to specific systems, such as Global Distribution Systems, Hotel Management Systems, and Accounting and Financial Management Systems.

Table 9. Rate your familiarity with the following information systems

		Very unfamiliar = 1		Unfamiliar = 2		Neither familiar nor unfamiliar = 3		Familiar = 4		Very familiar = 5		N	mean
		n1	%	n2	%	n3	%	n4	%	n5	%		
1	Global Distribution System	17	12.8%	14	10.5%	22	16.5%	54	40.6%	26	19.5%	**133**	3.44
2	Hotel Management Systems	35	26.3%	35	26.3%	32	24.1%	25	18.8%	6	4.5%	**133**	2.49
3	Accounting and Financial Management Systems	15	11.3%	27	20.3%	27	20.3%	45	33.8%	19	14.3%	**133**	3.20

(Source: Krousaraki, 2024).

The findings in Table 9 indicate that participants are more familiar with Global Distribution Systems, with the majority of responses (60.2%, or 80 out of 133 participants) indicating that they are "Familiar" or "Very familiar". In contrast, there is reduced familiarity with Hotel Management Systems, particularly among employees who are not directly involved in hotel booking processes (23.3%, or 31 participants). Additionally, Accounting and Financial Management Systems are more relevant to employees in financial departments and administrative positions, where familiarity with their use is higher (48.1%, or 64 participants).

The **second** question concerns the importance of criteria for selecting an information system by a Tour Operator. Participants are asked to evaluate how important they consider the thirteen criteria listed in the questionnaire. These criteria include available functionalities, the adoption of new technology, acquisition cost, price-quality ratio, system prestige, usability, training provision, availability of training materials, continuous technical support and maintenance, continuous updating, as well as the number and brands of other Tour Operators using the system. Participants are asked to rate the importance of these criteria for selecting an information system.

Table 10. Importance of the criteria for selecting an information system by a Tour Operator

		Very unimportant = 1		Unimportant = 2		Neutral = 3		Important = 4		Very important = 5		N	mean
		n1	%	n2	%	n3	%	n4	%	n5	%		
1	Available functionalities	1	0.8%	0	0.0%	4	3.0%	43	32.3%	85	63.9%	**133**	4.59
2	New technology	1	0.8%	1	0.8%	18	13.5%	61	45.9%	52	39.1%	**133**	4.22
3	Acquisition cost	1	0.8%	4	3.0%	29	21.8%	63	47.4%	36	27.1%	**133**	3.97
4	Price-quality ratio	1	0.8%	1	0.8%	18	13.5%	61	45.9%	52	39.1%	**133**	4.22
5	System prestige	5	3.8%	12	9.0%	49	36.8%	48	36.1%	19	14.3%	**133**	3.48
6	System usability	1	0.8%	1	0.8%	9	6.8%	47	35.3%	75	56.4%	**133**	4.46
7	Training provision	1	0.8%	1	0.8%	14	10.5%	59	44.4%	58	43.6%	**133**	4.29
8	Availability of training material	1	0.8%	2	1.5%	12	9.0%	65	48.9%	53	39.8%	**133**	4.26
9	Continuous technical support (e.g. phone support)	1	0.8%	2	1.5%	6	4.5%	53	39.8%	71	53.4%	**133**	4.44
10	Continuous maintenance (e.g. technician visits)	1	0.8%	2	1.5%	15	11.3%	53	39.8%	62	46.6%	**133**	4.30
11	Continuous upgrading (e.g. new releases)	1	0.8%	2	1.5%	8	6.0%	60	45.1%	62	46.6%	**133**	4.35
12	Number of other Tour Operators using it	5	3.8%	13	9.8%	45	33.8%	55	41.4%	15	11.3%	**133**	3.47

(*continued*)

Table 10. (*continued*)

		Very unimportant = 1		Unimportant = 2		Neutral = 3		Important = 4		Very important = 5		N	mean
		n1	%	n2	%	n3	%	n4	%	n5	%		
13	Brands of other Tour Operators using it	6	4.5%	13	9.8%	49	36.8%	52	39.1%	13	9.8%	**133**	3.40

(Source: Krousaraki, 2024).

The conclusions drawn from the analysis of the data in Table 10 are as follows: Available functionalities (1) are considered the most important criterion, with an average score (M.O.) of 4.59, with the majority of responses (63.9%, 85 participants) rating it as "Very important". The system's usability (6) and continuous technical support (9) are also very important, with average scores of 4.46 and 4.44, respectively. Criteria related to training provision (7), maintenance (10), and upgrades (11) have average scores ranging from 4.29 to 4.35, indicating their high importance.

The number and the brands of other Tour Operators (12 & 13) using the system are considered less important, with average scores of 3.47 and 3.40, highlighting the influence of these brands but not as a dominant characteristic. Participants rate as very important the functionality (M.O. 4.59), usability (M.O. 4.46), and continuous technical support (M.O. 4.44), as evidenced by the highest average scores in these criteria. This indicates that these factors have a direct impact on the efficiency and satisfaction of information system users in performing their tasks. The system's prestige (M.O. 3.48), the number (M.O. 3.47), and the brands of other Tour Operators using it (M.O. 3.40) range from "Neutral" to "Important", emphasizing the influence of these brands but not as a dominant characteristic.

The conclusion from the data analysis shows that criteria related to functionality, usability, and continuous technical support are the most important factors for information system users. The high average scores in these criteria underline their direct impact on user efficiency and satisfaction. In contrast, the number and brands of other Tour Operators using the system are considered less important factors, which may have greater weight for management rather than end users. Overall, the emphasis is on the practical utility and support provided by the system.

The **third** question concerns the importance of the benefits of using information systems in Tour Operator businesses. Participants are asked to evaluate the importance of nine benefits derived from the use of these systems. The benefits include: more effective marketing strategies, business promotion, customer approach, increase in sales, increase in customer satisfaction, developing new products, enhanced presence on digital distribution channels, collaboration to develop tourism packages, and collaboration for reservation management.

The conclusions drawn from the analysis of the data in Table 11 are as follows: The increase in customer satisfaction (5) and the increase in sales (4) emerge as the most significant benefits of information systems, with average scores of 4.43 and 4.39,

Table 11. Importance of the benefits of using information systems in Tour Operator businesses

		Very unimportant = 1		Unimportant = 2		Neutral = 3		Important = 4		Very important = 5		N	mean
		n1	%	n2	%	n3	%	n4	%	n5	%		
1	More effective marketing strategies	2	1.5%	3	2.3%	36	27.1%	64	48.1%	28	21.1%	**133**	3.85
2	Business promotion	1	0.8%	3	2.3%	43	32.3%	60	45.1%	26	19.5%	**133**	3.80
3	Customer approach	1	0.8%	2	1.5%	19	14.3%	55	41.4%	56	42.1%	**133**	4.23
4	Increase in sales	1	0.8%	2	1.5%	11	8.3%	49	36.8%	70	52.6%	**133**	4.39
5	Increase in customer satisfaction	1	0.8%	1	0.8%	11	8.3%	47	35.3%	73	54.9%	**133**	4.43
6	Developing new products that meet market demands	1	0.8%	2	1.5%	10	7.5%	69	51.9%	51	38.3%	**133**	4.26
7	Enhanced presence on digital distribution channels	1	0.8%	0	0.0%	16	12.0%	64	48.1%	52	39.1%	**133**	4.25
8	Collaboration to develop tourism packages (tourism packages offering more options to travelers)	1	0.8%	1	0.8%	17	12.8%	63	47.4%	51	38.3%	**133**	4.22
9	Collaboration for reservation management (Increasing customer experience)	1	0.8%	1	0.8%	18	13.5%	56	42.1%	57	42.9%	**133**	4.26

(Source: Krousaraki, 2024).

respectively. Specifically, 73 participants (54.9%) consider the increase in customer satisfaction "Very important", while 70 participants (52.6%) consider the increase in sales "Very important". These results indicate that employees in Tour Operators place great emphasis on both customer satisfaction and sales growth, recognizing that customer satisfaction and trust are critical for long-term success, while sales growth is vital for business development and profitability.

The development of new products that meet market demands (6) and the collaboration for reservation management (9) also have high average scores of 4.26. Specifically, 69 participants (51.9%) consider the creation of new products "Important", while 57 participants (42.9%) consider the collaboration for reservation management "Very important". These results show that employees in Tour Operators recognize the value of innovation and effective booking management for improving customer experience.

Enhanced presence in digital distribution channels (7) and customer approach (3) have average scores of 4.25 and 4.23, respectively. Enhanced presence in digital distribution channels is considered "Important" by 64 participants (48.1%), while customer approach is considered "Very important" by 56 participants (42.1%). These figures highlight the importance of digital presence and direct communication with customers for the success of Tour Operators.

Finally, more effective marketing strategies (1) and business promotion (2) are considered less important, with average scores of 3.85 and 3.80, respectively. 36 participants (27.1%) consider effective marketing strategies "Neutral", while 43 participants (32.3%) consider business promotion "Neutral". This suggests that, while participants recognize the importance of these aspects, they consider them less critical compared to other functions of information systems.

Overall, participants place greater emphasis on benefits directly related to sales growth, customer satisfaction, and the improvement of business processes, rather than aspects related to promotion and marketing strategies. These findings point to the need for continuous improvement of information systems and the development of strategies that focus on customer satisfaction and sales growth, highlighting the importance of innovation and effective management for the long-term success of Tour Operators.

The **fourth** question concerns the problems arising from the use of information systems in Tour Operator businesses. Participants are asked to evaluate the importance of six potential problems that may arise from the use of these systems. The problems include: difficulty adapting to the continues changes of the digital environment, difficulty monitoring changes in traveler preferences, difficulty providing personalized experiences, difficulty adapting products and services to new needs, difficulty adjusting marketing strategies, and ineffective management of distribution channels.

The conclusions drawn from the analysis of the data in Table 12 indicate that employees in Tour Operators consider the adjusting products and services to the new needs of travelers (4) as the most serious problem, with this difficulty having the highest average score (4.14). Specifically, 48.1% (64 participants) consider this difficulty "Important", while 34.6% (46 participants) consider it "Very Important". Providing personalized experiences to travelers (3) is also critical, with an average score of 4.02, as 47.4% (63 participants) consider it "Important" and 30.1% (40 participants) consider it "Very Important". Ineffective management of distribution channels (6) has an average score of 3.99, with 51.9% (69 participants) considering it "Important" and 25.6% (34 participants) considering it as "Very Important". The difficulty monitoring changes in traveler preferences (2) has an average score of 3.89, with 54.1% of participants (72 participants) considering it "Important" and 21.1% (28 participants) considering it "Neutral".

Table 12. Importance of problems arising of using information systems in Tour Operator businesses

		Very unimportant = 1		Unimportant = 2		Neutral = 3		Important = 4		Very important = 5		N	mean
		n1	%	n2	%	n3	%	n4	%	n5	%		
1	Difficulty adapting to the continuous changes of the digital environment	3	2.3%	9	6.8%	24	18.0%	69	51.9%	28	21.1%	**133**	3.83
2	Difficulty monitoring changes in traveler preferences	1	0.8%	5	3.8%	28	21.1%	72	54.1%	27	20.3%	**133**	3.89
3	Difficulty providing personalized experiences to travelers	2	1.5%	4	3.0%	24	18.0%	63	47.4%	40	30.1%	**133**	4.02
4	Difficulty adjusting products/services to the new needs of travelers	1	0.8%	3	2.3%	19	14.3%	64	48.1%	46	34.6%	**133**	4.14
5	Difficulty adapting marketing strategies	1	0.8%	7	5.3%	33	24.8%	65	48.9%	27	20.3%	**133**	3.83
6	Ineffective management of distribution channels	1	0.8%	3	2.3%	26	19.5%	69	51.9%	34	25.6%	**133**	3.99

(Source: Krousaraki, 2024).

Additionally, continuous changes in the digital environment (1) have an average score of 3.83, with 51.9% (69 participants) considering this difficulty "Important" and 21.1% (28 participants) considering it "Very Important". The adaptation of marketing strategies (5) also has an average score of 3.83, with 48.9% (65 participants) considering it as "Important" and 24.8% (33 participants) considering it "Neutral". Overall, the data indicate that employees in Tour Operators prioritize strategies that improve customer experience and increase their competitiveness in the market.

The survey participants seem to recognize the limitations of current information systems in providing personalized experiences and adapting services to the constantly changing needs of travelers. They likely believe that, despite technological advancements, the human factor remains crucial in achieving these goals. However, with the evolution of Artificial Intelligence (AI), it is expected that information systems will acquire advanced capabilities that will allow them to offer these personalized experiences with greater accuracy and flexibility, thereby enhancing their value to Tour Operators.

5.4 Cross-Tabulation of Variables

In the context of this research on information systems for Tour Operators and based on the need for further analysis of the findings, it was decided to conduct cross-tabulations focusing on marketing functions, such as strategies and promotion, in relation to the departments of a Tour Operator business and the importance of the benefits as perceived by users and employees in these departments. These cross-tabulations were chosen to highlight the importance of user experience, providing valuable insights into the actual capabilities of information systems (Table 13).

Table 13. More effective marketing strategies/Job position in the company

More effective marketing strategies											
Job Position in the Company	Very unimportant		Unimportant		Neutral		Important		Very important		N
	n1	%	n2	%	n3	%	n4	%	n5	%	
Category Management Department	0	0.0%	2	13.3%	3	20.0%	7	46.7%	3	20.0%	15
Customer Service Department	0	0.0%	0	0.0%	0	0.0%	0	0.0%	1	100.0%	1
Finance, Accounting and HR Department	0	0.0%	0	0.0%	6	37.5%	9	56.3%	1	6.3%	16
Operations Department	1	3.2%	0	0.0%	9	29.0%	13	41.9%	8	25.8%	31
Operations Support	1	2.1%	1	2.1%	11	22.9%	26	54.2%	9	18.8%	48
Product Development and Planning Department	0	0.0%	0	0.0%	4	36.4%	5	45.5%	2	18.2%	11
Sales and Marketing Department	0	0.0%	0	0.0%	0	0.0%	1	25.0%	3	75.0%	4

(continued)

Table 13. (*continued*)

More effective marketing strategies											
Job Position in the Company	Very unimportant		Unimportant		Neutral		Important		Very important		N
	n1	%	n2	%	n3	%	n4	%	n5	%	
Technology and IT Department	0	0.0%	0	0.0%	3	42.9%	3	42.9%	1	14.3%	7
	2		3		36		64		28		133

(Source: Krousaraki, 2024).

The majority of participants from all departments evaluated the benefits of using information systems for more effective marketing strategies, ranging from "Neutral" to "Important". More specifically, the job positions, within the company, provided us with the most characteristic responses in relation to the number of employees who have been recorded. Consequently, our focus is on the Finance, Accounting, and HR Departments, Operations Support, and Sales and Marketing Departments. The analysis focuses on departments that demonstrated higher scores, as these results are most pertinent to the central themes of the study and provide valuable insights.

For the Finance, Accounting, and HR Department (56.3%, 9 out of 16 participants) and the Operations Support (54.2%, 26 out of 48 participants), these benefits are considered "Important". Meanwhile, for 75% (3 out of 4 participants) of the Sales and Marketing Department, it is naturally deemed an "Very important" function. Participants likely believe that the business should develop specialized and focused marketing strategies independently of information systems (Table 14).

Table 14. Business promotion/Job position in the company

Business promotion											
Job Position in the Company	Very unimportant		Unimportant		Neutral		Important		Very important		N
	n1	%	n2	%	n3	%	n4	%	n5	%	
Category Management Department	0	0.0%	1	6.7%	8	53.3%	5	33.3%	1	6.7%	15
Customer Service Department	0	0.0%	0	0.0%	0	0.0%	0	0.0%	1	100.0%	1

(*continued*)

Table 14. (*continued*)

Business promotion											
Job Position in the Company	Very unimportant		Unimportant		Neutral		Important		Very important		N
	n1	%	n2	%	n3	%	n4	%	n5	%	
Finance, Accounting and HR Department	0	0.0%	0	0.0%	8	50.0%	7	43.8%	1	6.3%	16
Operations Department	1	3.2%	0	0.0%	10	32.3%	12	38.7%	8	25.8%	31
Operations Support	0	0.0%	2	4.2%	12	25.0%	20	41.7%	14	29.2%	48
Product Development and Planning Department	0	0.0%	0	0.0%	3	27.3%	8	72.7%	0	0.0%	11
Sales and Marketing Department	0	0.0%	0	0.0%	1	25.0%	2	50.0%	1	25.0%	4
Technology and IT Department	0	0.0%	0	0.0%	1	14.3%	6	85.7%	0	0.0%	7
	1		3		43		60		26		133

(Source: Krousaraki, 2024).

The majority of participants, across all departments, evaluated the benefits of using information systems for business promotion as ranging from "Neutral" to "Important". More specifically, the job positions that provide the most representative responses in relation to the number of employees are recorded. Consequently, our focus is on the Finance, Accounting, and HR Departments, Category Management, as well as Product Development and Planning departments. The analysis focuses on these departments, as the higher scores they presented are directly related to the main issues of the research, providing valuable insights into the study.

For the Finance, Accounting, and HR Department (50%, 8 out of 16 participants) and the Category Management (53,3%, 8 out of 15 participants), these benefits are considered "Neutral". Meanwhile, 72.7% (8 out of 11 participants) of the Product Development and Planning department, it is deemed "Important" function. Participants may believe that the business should develop other marketing strategies for promotion independently of information systems.

These findings confirm the theory regarding the value of information systems, while simultaneously highlighting the fact that they cannot replace the autonomous functions and strategies of marketing and the implementation of these strategies by specialized marketing personnel (Table 15).

Table 15. Difficulty adjusting products/services to the new needs of travelers/ Job position in the company

Difficulty adjusting products/services to the new needs of travelers											
Job Position in the Company	Very unimportant		Unimportant		Neutral		Important		Very important		N
	n1	%	n2	%	n3	%	n4	%	n5	%	
Category Management Department	0	0.0%	0	0.0%	2	13.3%	7	46.7%	6	40.0%	15
Customer Service Department	0	0.0%	0	0.0%	0	0.0%	0	0.0%	1	100.0%	1
Finance, Accounting and HR Department	0	0.0%	0	0.0%	3	18.8%	10	62.5%	3	18.8%	16
Operations Department	1	3.2%	0	0.0%	4	12.9%	15	48.4%	11	35.5%	31
Operations Support	0	0.0%	2	4.2%	7	14.6%	21	43.8%	18	37.5%	48
Product Development and Planning Department	0	0.0%	1	9.1%	1	9.1%	6	54.5%	3	27.3%	11
Sales and Marketing Department	0	0.0%	0	0.0%	0	0.0%	1	25.0%	3	75.0%	4
Technology and IT Department	0	0.0%	0	0.0%	2	28.6%	4	57.1%	1	14.3%	7
	1		3		19		64		46		133

(Source: Krousaraki, 2024).

These findings indicate that the difficulty in adapting products/services to the new needs of travelers is "Important" to "Very important" problem for the majority of all departments. Although there are only 4 participants from the Sales and Marketing departments, the response of 75% of them, rating it as "Very important", suggests that information systems have limited capabilities in providing personalized experiences and adapting services to the constantly changing needs of travelers.

Additionally, this finding confirms the theory regarding the concerns but also shows that user experience can add knowledge about the actual capabilities and the necessity of the human factor in offering personalized and comprehensive services.

6 Conclusions

Through a review of the literature, the importance of utilizing information systems by Tour Operators is confirmed. Particular emphasis is placed on the use of Global Distribution Systems (GDS), which are required not only to meet the operational needs of businesses but also to adapt to the ever-changing demands of the tourism market. The literature highlights the advantages of leveraging advanced information systems, such as the improvement of customer experience, the enhancement of the visibility of tourism enterprises, the attraction of new customers, and the assurance of high-quality services in the digital tourism environment. Additionally, the contribution of information systems to the creation of innovative products is emphasized, while the challenges concerning the adaptation of tourism enterprises to advanced technologies and the requirements of contemporary marketing are also noted.

The research confirms the advantages offered by the use of information systems. However, this study at the same time achieves an evaluation based on the importance attributed to these systems by the users themselves, which constitutes an additional contribution of the study to both the literature and the professional field of technology and T.O. marketing. Furthermore, this study, always from the user's perspective, manages to identify concerns and weaknesses, once again evaluating them based on the importance attributed to them by the users, thereby providing knowledge to both the scientific literature and the professional field of technology and T.O. marketing.

The sample of the present study includes a significant number of employees with extensive work experience and esteemed positions in the field of Tour Operating. This enhances the validity and quality of the results, as the conclusions are based on the opinions and perceptions of a dynamic group with deep knowledge and experience. Additionally, the high educational level of the sample, with the majority having tertiary education and a significant proportion holding postgraduate degrees, indicates that the participants possess the necessary theoretical and practical knowledge to provide valid and targeted opinions. The contribution of employees specialized in tourism is also critical, as they add depth to the understanding of tourism products and markets. Finally, the participation from the Operations and Operations Support departments demonstrates strong representation in an area crucial for the functioning and success of Tour Operators, further adding weight to the study's findings.

The conclusions derived from the research findings are summarized as follows:

1. Familiarity with systems: Overall, employees in Tour Operators appear to be more familiar with Global Distribution Systems and Accounting management systems compared to Hotel management systems. This is likely because those working in transportation and excursion departments do not frequently need to manage such systems.
2. Limitations in personalizing experiences: Employees in Tour Operators believe that current information systems present limitations in providing personalized experiences

to travelers. Despite technological advancements, the human factor is still considered irreplaceable for achieving this goal. However, with the evolution of Artificial Intelligence (AI), systems are expected to acquire more advanced capabilities that will allow for greater accuracy and flexibility in providing personalized services, thereby enhancing their value to Tour Operators.

3. Recognition of limited contribution to success: Despite the use of information systems, employees in Tour Operators express concerns about the overall success of the company's tourism activities. This may indicate that users believe these systems have not yet reached the level needed to fully address the industry's challenges. Therefore, there is a need for continuous improvement of the systems to contribute significantly to the company's success.
4. Regarding the issue of promotion: Employees in Tour Operators may consider that the company needs to develop specialized and focused marketing strategies that do not rely solely on information systems. This suggests that the promotion of services may require a more holistic and targeted approach beyond the simple use of technology.

Summarizing the research findings, it becomes clear that employees in Tour Operators identify obstacles in the utilization of information systems, particularly in terms of personalizing experiences and contributing to the success of tourism activities. Despite technological advancements, human intervention is still deemed necessary. Employees note that the existing systems do not fully meet the industry's needs, emphasizing the necessity for continuous upgrades and development. Additionally, they believe that Tour Operators need to develop targeted marketing strategies that integrate both technology and the human element for better results.

With the development of Artificial Intelligence (AI), information systems are expected to acquire more advanced capabilities, allowing for greater flexibility and accuracy in personalizing services, thereby increasing their value in the Tour Operators sector. AI may need to be considered as the tool that will contribute to managing user feedback and observations and solving various problems.

Future research, due to the significance of Artificial Intelligence (AI), should focus on its application in the Tour Operators sector. Studying best practices for integrating AI technologies could provide valuable insights for enhancing the traveler experience and the strategic development of the industry.

Despite the increasing use of technology, the human role remains crucial for providing quality and personalized services to travelers.

References

1. Krousaraki, A.: The impact of information systems in the management and marketing of tour operators: challenges and prospects (Unpublished master's thesis). Hellenic Open University (2024)
2. Yarcan, S., Cetin, G.: Tour Operating Business. Istanbul University Press (2021). https://doi.org/10.26650/B/SS23.2021.015
3. Buhalis, D.: Strategic use of information in the tourism industry. Tour. Manag. **19**(5), 409–421 (1998)

4. Zekanović-Korona, L., Grzunov, J.: Facilitating smart tourism destination development: towards a model of information system for smart tourism. Varazdin: Faculty of Organization and Informatics Varazdin (2022). https://proxy.eap.gr/login?url=. https://www.proquest.com/conference-papers-proceedings/facilitating-smart-tourism-destination/docview/2720979401/se-2
5. Olive, A.: Conceptual Modeling of Information Systems (1; 2007; ed.). Springer Nature (2007). https://doi.org/10.1007/978-3-540-39390-0
6. Moutzouri, V.: Information Systems in Tourism (2012). https://dione.lib.unipi.gr/xmlui/bitstream/handle/unipi/5105/Moutzouri.pdf?seq. Accessed 17 Jan 2024
7. Benefits of Global Distribution System I GDS Booking System. (n.d.). www.flightslogic.com. https://www.flightslogic.com/benefits-of-global-distribution-system.php. Accessed 17 Jan 2024
8. Best marketing tools for tourism organizations, 28 September 2022. MySmartJourney. https://mysmartjourney.com/en-ca/post/7-best-marketing-tools-for-tourism-organizations. Accessed 24 Feb 2024
9. Kotler, P., Bowen, J., Makens, J., Baloglu, S.: Marketing Tourism & Hospitality, pp. 30–49. Broken Hill Publishers Ltd. (2019)
10. Bugembe, D.: How tour operators will benefit from digitalised tourism activities. Daily Monitor (2022). https://proxy.eap.gr/login?url=. https://www.proquest.com/newspapers/how-tour-operators-will-benefit-digitalised/docview/2755108395/se-2. Accessed 22 Jan 2024
11. Rather, R.A.: Customer experience and engagement in tourism destinations: the experiential marketing perspective. J. Travel Tour. Mark. **37**(1), 15–32 (2020). https://doi.org/10.1080/10548408.2019.1686101
12. Hollebeek, L.D., Macky, K.: Digital content marketing's role in fostering consumer engagement, trust, and value: framework, fundamental propositions, and implications. J. Interact. Mark. **45**, 27–41 (2019)
13. Hollebeek, L.D., Glynn, M.S., Brodie, R.J.: Consumer brand engagement in social media: conceptualization, scale development and validation. J. Interact. Mark. **28**(2), 149–165 (2014)
14. Hultman, M., Skarmeas, D., Oghazi, P., Beheshti, H.M.: Achieving tourist loyalty through destination personality, satisfaction, and identification. J. Bus. Res. **68**(11), 2227–2231 (2015)
15. Kumar, V., Rajan, B., Gupta, S., Dalla Pozza, I.: Customer engagement in service. J. Acad. Mark. Sci. **47**(1), 138–160 (2019)
16. Lemon, K.N., Verhoef, P.C.: Understanding customer experience throughout the customer journey. J. Mark. **80**(6), 69–96 (2016)
17. Pine, B.J., Gilmore, J.H.: Welcome to the experience economy. Harv. Bus. Rev. **76**(4), 97–105 (1998)
18. Prahalad, C.K., Ramaswamy, V.: The new frontier of experience innovation. MIT Sloan Manag. Rev. **44**(4), 12–18 (2003)
19. Usakli, A., Baloglu, S.: Brand personality of tourist destinations: an application of self-congruity theory. Tour. Manag. **32**(1), 114–127 (2011)
20. Braimah, S.M., Solomon, E.N.A., Hinson, R.E.: Tourists satisfaction in destination selection determinants and revisit intentions; perspectives from Ghana. Cogent Soc. Sci. **10**(1) (2024). https://doi.org/10.1080/23311886.2024.2318864
21. Pencarelli, T.: The digital revolution in the travel and tourism industry. Inf. Technol. Tourism **22**(3), 455–476 (2020). https://doi.org/10.1007/s40558-019-00160-3
22. Abualloush, S., Bataineh, K., Aladwan, A.S.: Impact of information systems on innovation (product innovation, process innovation) - field study on the housing bank in Jordon. Int. J. Bus. Adm. **8**(1), 95 (2016). https://doi.org/10.5430/ijba.v8n1p95
23. Blana, M.: The Use of New Technologies and Digital Tools in Tourism. TravelDailyNews Greece & Cyprus (2023). https://www.traveldailynews.gr/technologia/chrisi-neon-technologion-kai-psifiakonergaleion-ston-toyrismo/. Accessed 03 Mar 2024

24. Wilson, K.: 10 Innovative Tourism Product Ideas Using AI-Powered Technology - Leve Global (2023). https://leve-global.com/10-innovative-tourism-product-ideas-using-ai-powered-technology/. Accessed 31 Mar 2024
25. Bilgihan, A., Bujisic, B.: The effect of website features in online relationship marketing: a case of online hotel booking. Electron. Commer. Res. Appl. **14**(4), 222–232 (2015). https://doi.org/10.1016/j.elerap.2014.09.001. ISSN 1567-4223
26. What are the benefits and challenges of using digital platforms as distribution channels? (n.d.). www.linkedin.com. Powered by AI and the LinkedIn community. https://www.linkedin.com/advice/0/what-benefits-challenges-using-digital-platforms-1f
27. Buhalis, D., Amaranggana, A.: Smart tourism destinations. In: Information and Communication Technologies in Tourism 2014, pp 553–564. Springer, Cham (2013). https://doi.org/10.1007/978-3-319-03973-2_40
28. Buhalis, D., Foerste, M.: SoCoMo marketing for travel and tourism: empowering co-creation of value. J. Destin Mark. Manag. **4**(3), 151–161 (2015). https://doi.org/10.1016/j.jdmm.2015.04.001
29. Gretzel, U., Werthner, H., Koo, C., Lamsfus, C.: Conceptual foundations for understanding smart tourism ecosystems. Comput. Hum. Behav. **50**, 558–563 (2015). https://doi.org/10.1016/j.chb.2015.03.043
30. Koo, C., Shin, S., Gretzel, U., Hunter, W.C., Chung, N.: Conceptualization of smart tourism destination competitiveness. Asia Pac. J. Inf. Syst. **26**(4), 561–576 (2016). https://doi.org/10.14329/apjis.2016.26.4.561
31. Li, Y., Hu, C., Huang, C., Duan, L.: The concept of smart tourism in the context of tourism information services. Tour. Manag. **58**, 293–300 (2017). https://doi.org/10.1016/j.tourman.2016.03.014
32. Segittur. Report on smart destinations. Building the future (2015). https://www.segittur.es/opencms/export/sites/segitur/content/galerias/descargas/documentos-en/Smart-Destination.pdf
33. Xiang, Z., Tussyadiah, I., Buhalis, D.: Smart destinations: foundations, analytics, and applications. J. Destin. Mark. Manag. **4**(3), 143–144 (2015). https://doi.org/10.1007/s12525-015-0196-8

Exploring the Impact of Artificial Intelligence on the Tourism Industry: Opportunities and Challenges

Ioannis Katsanakis[1], Georgia Zouni[1(✉)], Stergios Spasos[2], Dimitra-Margieta Lykoudi[1], and Nikolaos Akriotis[1]

[1] Tourism Studies, University of Piraeus, 18534 Piraeus, Greece
gzouni@unipi.gr

[2] MBA Tourism Management, University of Piraeus, 18534 Piraeus, Greece

Abstract. Technological innovations have significantly changed the tourism industry over the past few years, with artificial intelligence (AI) emerging as a transformative technology capable to revolutionize the industry. This study examines the opportunities and challenges of integrating AI into the tourism sector. In particular, it examines the effects of AI on improving customer experience, streamlining operations and providing personalized travel solutions, alongside challenges such as privacy, ethics and workforce adaptation. This study, using qualitative research and semi-structured interviews, tries to uncover the opinions and thoughts of travel and tourism professionals on how AI has influenced the way they work and operate. Applying thematic analysis six key themes have been identified: process automation, perceptions of AI, future industry direction, ethical considerations, workforce training, and strategic decision-making. Findings reveal that although AI can improve decision-making, reduce operational costs, and enhance service personalization, its integration raises concerns about job displacement, data security, and ethical practices. The study highlights the need to address the ethical and legal implications of AI in the tourism sector to ensure its responsible adoption. Furthermore, it underlines the importance of lifelong learning and workforce reskilling to prepare professionals for the evolving demands of AI-driven tourism industry. This research contributes to the growing literature on AI in tourism by offering a focused investigation of its real-world applications and impacts and providing actionable insights for industry stakeholders seeking to leverage AI to gain competitive advantage while addressing potential risks. Future research should include an in-depth study of customer perceptions of AI-driven services and the development of strategies to ensure harmonious human-technology collaboration in tourism.

Keywords: Artificial Intelligence · Tourism Industry · Thematic Analysis

A. K. Zarkada (Ed.): GMA-GAMMA 2025, SPBE, pp. 154–178, 2025.
https://doi.org/10.1007/978-3-031-96930-0_9

1 Introduction

Artificial intelligence (AI) is a cutting-edge technology that has revolutionized our daily lives. Artificial intelligence has made significant strides in recent years, from virtual assistants in our smartphones to complex decision-making systems. The tourism industry is no exception, with AI applications now embedded in diverse areas, from customer service to operational optimization. However, to understand AI's potential in this field, it is essential to clarify the concept of intelligence and its artificial counterpart. Intelligence is a collection of abilities, including the ability to comprehend the surrounding environment, leverage past experiences, and integrate existing knowledge to effectively address new challenges [1]. Gretzel [2] further elaborates on these attributes, stating that "intelligent" systems possess the capacity to interpret their environment, self-train, and apply acquired knowledge to future scenarios. A defining characteristic of artificial intelligence is its capability to execute tasks that were traditionally regarded as exclusively human. This is made possible through a range of cognitive functions, such as perception, learning, abstraction, reasoning, and communication [3].

According to Brooks [4], artificial intelligence systems are classified according to the range of capabilities they have and characterized respectively as "narrow" or "general". While current AI systems are categorized as "narrow AI" designed to perform specific tasks, their applications in tourism - such as recommendation systems, chatbots, and predictive analytics - demonstrate significant potential for enhancing customer experiences and operational efficiency [5].

In general, there is a wide variety of techniques and subfields that contribute to the development of artificial intelligence systems, such as:

a) *Machine Learning* (ML): This is at the heart of artificial intelligence, allowing systems to learn from data and improve their performance. Some machine learning applications concern the health system, which is related to the diagnosis and prediction of treatment results, the financial sector to investigate financial fraud, means of transportation to find the fastest route, etc. [6]. As Sterne [7] points out, machine learning is designed to acquire knowledge rather than follow instructions. Specifically, it involves training machines to perform tasks once done exclusively by humans. Understanding patterns like this is important for machine learning to effectively utilize its knowledge in solving organizational issues that will arise. Unlike traditional methods that rely solely on data, these systems can adapt and modify their perspectives based on the information they encounter. The use of artificial intelligence and machine learning in the hotel industry, as highlighted by Millauer & Vellekoop [8] and Viverit et al. [9], allows a more accurate prediction of occupancy rates. This data-driven approach allows hotels to optimize room rates by considering various factors, including future supply and demand.
b) *Robotics*: This is an interdisciplinary field that combines artificial intelligence with mechanical and electrical engineering to design, build, and operate robots that can perform a variety of tasks autonomously or semi-autonomously [10].
c) *Deep Learning* (DL): This is a more specialized level of machine learning that helps solve problems requiring complex recognition and data understanding. Its applications are widespread in various scientific disciplines, such as computer vision, natural language processing, which allows computers to understand human language,

and speech recognition, which allows the conversion of spoken speech into text for computer processing [11].

d) *Natural Language Processing* (NLP): This is one of the many applications of engineering and deep learning aimed at recognizing the human voice. Characteristic examples are chatbots and virtual assistants, which have the ability to communicate with other people using human language (with practical applications in customer service, health, education, machines that translate texts, etc.) [12]. The data that AI systems process is derived from statistical correlation between large amounts of information, allowing machines to perform tasks that traditionally require human intelligence. AI is a new form of intelligence that differs from human intelligence because it is data-driven and operates using non-human methods.

Although artificial intelligence has experienced rapid growth in technological development, predicting its future is difficult due to the speed at which developments occur. However, the potential for further development of artificial intelligence remains high. We are currently in the era of "weak" artificial intelligence, as machines are able to perform tasks just like humans but much faster. Artificial general intelligence, or vital artificial intelligence, which was the original goal of artificial intelligence, has not yet been achieved [13]. Recent research has shifted its focus from simply evaluating the effectiveness of artificial intelligence to testing its effects and outcomes. This is a notable departure from previous years when research focused primarily on AI performance [14].

Despite these benefits, integrating AI into tourism raises several challenges. Issues such as privacy, security, workforce adaptation, and ethical considerations require careful evaluation. This study addresses these challenges and explores AI's dual role as a disruptor and enabler within the tourism industry. By examining its applications, benefits, and risks, this research aims to provide a nuanced understanding of AI's transformative impact and its implications for the future of tourism.

2 Applications of Artificial Intelligence in the Tourism Sector

AI has great potential to significantly impact the hospitality and tourism sectors by enhancing operational performance and improving customer service, ultimately leading to higher profitability [15, 16]. For example, both Melián-González et al. [17] and Yun & Park [18] highlight how AI-powered chatbots can be used to provide 24/7 interactive customer service, answer common guest questions, provide personalized recommendations, offers, and assistance, and even handle simple booking requests. This can improve customer service and reduce response times, enhancing guest engagement, loyalty and satisfaction [19].

AI is based on big data, processing capabilities, and algorithms. Each of these three elements has experienced leaps and bounds in recent years, as several trends have coincided: first, the refinement and advancement of artificial intelligence algorithms - second, significant improvements in data processing capabilities, and finally, in the context of big data, the development of new and more robust information sources and architectures that enable the seamless storage and processing of massive amounts of data. These improvements have, in turn, fueled significant improvements in AI systems and robotics, known as the fourth industrial revolution [20]. Currently, AI applications are being designed and implemented across various aspects of the travel and tourism industry. These include personalization and recommendation systems, virtual travel assistants, chatbots, predictive analytics, machine translation tools, as well as voice recognition and natural language processing technologies.

Artificial intelligence plays a crucial role in the tourism sector for multiple reasons. Travelers must make numerous decisions regarding their trips, such as selecting a destination, choosing transportation methods, booking accommodations, and planning activities. These choices significantly influence overall travel satisfaction, making AI-driven solutions essential for enhancing decision-making and optimizing the tourist experience. However, the range of destinations, transport, accommodation and activities currently available to the visitor has a wide range that requires tools to make the best choice between them. Travel organizations and agents encounter a similar challenge in identifying the best alignment between customer preferences and suitable travel packages.

Matching demand with the right tourism offerings is a highly intricate process, one that artificial intelligence has the potential to optimize significantly. However, even after arriving at their destination, tourists must navigate unfamiliar environments, including different customs, languages, cultural norms, and local cuisine, among other aspects that may be new to them. At this point, artificial intelligence can help tourists improve their satisfaction with their needs through personalized proposals. Artificial intelligence simulates human intelligence processes by machines and special computer systems. It encompasses various technologies, including engineering, natural language processing and robotics. In particular, it already provides a wide range of capabilities not there before, such as a more streamlined experience for visitors, improved functions, and decision-making capabilities. Characteristically, AI-powered chatbots are now helping tourism businesses provide 24-h customer service, increase revenue potential and boost engagement. Additionally, artificial intelligence (AI) is recognized as a significant trend and advancement in the transportation industry, as highlighted in Deloitte's "Transportation Trends 2020" report. This development is expected to impact the global tourism market profoundly. Harnessing the power of data, analytics and cloud computing to enhance or alleviate traffic congestion, improve compliance and facilitate air traffic control [21].

The analysis of many studies and research on the applications of artificial intelligence in the tourism industry reveals that the penetration of artificial intelligence in every business sector has increased by 56% compared to 2019. This growth is also evident in the tourism sector, with artificial intelligence increasingly used to improve customer experiences, services and marketing strategies. Airlines and airports place particular emphasis on new technologies that offer them significant advantages both at a strategic and business level, as the safe and easy transportation of passengers and the use of biometric technology is one of their main priorities (use of autonomous vehicles, using big data analysis, use of robotics and blockchain). A typical example is Heathrow Airport in London, which incorporates electric self-driving machines to efficiently transport passengers along a designated route within the airport premises. British Airways uses the "Know Me" system to thoroughly analyze customer data, enabling it to provide personalized offers to travellers. San Jose Airport has successfully implemented robotic guides that help passengers reach their respective gates and entertain travellers through interactive activities such as dancing along or taking photos with them. Another example is the collaboration of the Government of Dubai with the innovative Blockchain start-up company "ObjectTech" to create digital passports using Blockchain technology. These passports aim to improve the passenger identification and verification process at Dubai International Airport [21].

All this leads to the conclusion that artificial intelligence (AI) is a beneficial element in the tourism industry, as evidenced by the fact that 84% of airlines and 61% of airports have intentions to either launch significant initiatives or allocate resources to research and development projects until the year 2021 that are directly related to AI. A typical example is KLM Royal Dutch Airlines, which pioneered such a program successfully. More specifically, to enhance the speed with which queries are processed on social media platforms, it is looking at incorporating artificial intelligence (AI). Heathrow Airport recently revealed plans to invest $63,000,000 in advanced AI security systems for facial recognition in airport terminals. Similarly, Singapore's Changi Airport equips its employees with state-of-the-art AR glasses, enabling them to access real-time instructions tailored to their needs [21].

Other studies have explored the personalized application of artificial intelligence in tourism. For example, Gaur et al. [22], conducted a study of a hotel chain that implemented an AI-based recommendation system to provide customers with a personalized booking process. Artificial intelligence algorithms analyze customer preferences, online behaviour, and other factors to provide personalized accommodations, activities, and destination recommendations. This system analyzes customer data, such as location and booking history, to develop hotel recommendations and special offers. This research concluded that AI systems significantly increase bookings and customer satisfaction. Prentice et al. [23], investigating the service of tourists through artificial satisfaction in the hotel industry, concluded that accommodations that used AI customer service had increased levels of customer satisfaction compared to those that did not use AI. Furthermore, they observed that the use of AI was positively related to customer loyalty.

Chatbot technology is one of the most popular artificial intelligence applications in the tourism industry. Chatbots use natural language processing (NLP) and AI-trained models to understand and answer user queries accurately. This technology has proven beneficial to travel businesses, reducing the need for human customer service representatives and saving time and money. More specifically, chatbots have the ability to answer customer queries, provide information, and even help with reservations, significantly reducing response times and increasing guest satisfaction levels. This is also confirmed by Ashfaq et al. [24], stating that chatbots can manage a considerable number of customer queries and provide appropriate answers in a short period of time, in turn cultivating high levels of satisfaction and loyalty.

In addition, AI through "predictive analysis" allows travel companies to make forecasts regarding demand, adjust prices and thus discover potential opportunities to grow further. To explain, "prediction" uses the technique of machine learning. Using specific data and statistical algorithms, it can predict a particular accommodation's demand, how a customer will behave and how to improve the advertising campaign. This particular analysis method is very effective for predicting occupancy and, by extension, the correct cost of accommodation services [25]. In addition, "forecasting" can be used to predict the visitation of tourist attractions [26], while it can also improve advertising campaigns even in a travel agency [27].

Artificial intelligence applications in the tourism sector are vast and constantly expanding. AI's potential benefits to improve customer experiences, understand traveller trends and behaviours, and revolutionize the entire industry are enormous. AI can be used to create personalized travel experiences, optimize processes, improve security, and facilitate check-ins, among other applications. More generally, integrating artificial intelligence in the tourism sector can also increase the reach of tourism products, thereby enhancing the promotion and marketing of the industry.

3 Benefits of Applying Artificial Intelligence (AI) in the Tourism Industry

There are many possibilities for using artificial intelligence in the field of tourism. The future of tourism, in general, will affect social and economic growth by integrating artificial intelligence. This can be achieved by developing a systematic knowledge generation process to ensure the beneficial application of intelligent automation in the tourism industry. The use of artificial intelligence in the tourism industry is just one aspect of a more significant technological revolution that has been underway for many years. The integration of artificial intelligence in the tourism industry has introduced several innovative concepts such as the "sharing economy", the "Internet of Things", and the "Internet of people" [28]. This has opened up new possibilities for tourism to leverage artificial intelligence to increase its profit at the tourism business level.

The development of artificial intelligence and its related technologies has also led to opportunities for intelligent automation in travel and tourism [29]. Using artificial intelligence, the tourism sector can achieve greater efficiency, and in the near future, personalized solutions for new options are expected to emerge [28]. In addition, the tourism industry already uses intelligent systems and chatbots powered by artificial intelligence [28], and it is expected that the application of intelligent automation in tourism will continue to increase [29]. The convergence of digital technologies and the tourism industry presents a unique opportunity for applying artificial intelligence. Modern systems can be used in the service sector to inform business practices and policies [29]. Therefore, research priorities should focus on creating beneficial AI, facilitating its adoption, assessing its impact, and ensuring sustainable development with AI [29]. This approach will guarantee the successful use of artificial intelligence in the tourism industry.

In the realm of operations, artificial intelligence optimizes resource allocation by assigning rooms and other facilities based on economic conditions and guest preferences. It enhances facility management by proactively overseeing maintenance, tailoring service offerings using historical data and tourist arrival forecasts, and customizing culinary options to align with customer tastes. AI also automates housekeeping through robotic cleaning systems, assists in evaluating and selecting suitable employees, and facilitates seamless interactions between guests and intelligent systems, such as during check-ins and service requests. Additionally, it enables the integration of real-time data into business processes, supports the deployment of robots at reception desks, as doormen, or for delivery services, and improves inventory and energy management. By leveraging AI, hospitality businesses can create a personalized and comfortable environment, grant guests access to digital services, and enhance financial management by factoring in projected revenues and tourist arrivals.

In marketing and merchandising, AI plays a critical role in refining forecasting, dynamically adjusting prices and promotional offers for current and prospective customers, and improving customer relationship management (CRM) systems. It facilitates the development of highly personalized services through mass customization and supports intelligent marketing strategies. AI-driven tools aid in generating tailored forecasts, assisting sales representatives, and deploying intelligent sales assistants. Furthermore, AI enables real-time, context-based marketing offers and enhances queue management systems to streamline customer interactions and improve overall service efficiency [17].

4 Enhancing the Customer Experience in the Tourism Industry Through Artificial Intelligence

AI technology is becoming increasingly significant in the travel industry, enabling businesses to deliver superior customer experiences through automation and personalization. AI reduces operational complexity while enhancing efficiency by streamlining customer service, ticketing, and payment tasks. Moreover, it identifies customer preferences, trends, and needs, enabling the delivery of highly tailored experiences [23]. AI can use advanced algorithms to predict guest preferences and provide customized recommendations, such as developing personalized tours and travel plans [30]. For example, AI-powered chatbots can interact with customers, provide 24/7 customer service and answer customer queries [31]. Additionally, AI can optimize marketing campaigns, reduce costs, and transform how travellers interact with businesses [32]. These technologies, from chatbots to predictive analytics, are revolutionizing the industry by creating more seamless and meaningful travel experiences. The following sections explore these AI-driven methods and tools in greater detail.

a) *Forecasting*: Forecasting is a predictive method that relies on historical and comparable data to estimate future trends, assisting decision-making across various sectors. It is widely used in industries requiring forward-looking strategies, including tourism. Forecasting is particularly well-suited for artificial intelligence (AI) algorithms [33], especially in the presence of big data, which enhances predictive accuracy [34]. AI-driven forecasting techniques are typically classified into five categories: grey theory, fuzzy time series, rough set approximation, support vector machines (SVM), and artificial neural networks (ANN) [33]. Within the tourism industry, forecasting plays a crucial role in understanding tourist demand [35], designing marketing strategies, managing financial resources and workforce allocation [33], identifying fraudulent activities in restaurants [36], and supporting facility management and maintenance needs [35].

   Despite its potential, AI-based forecasting should be applied cautiously due to mixed results in empirical studies. Yu and Schwartz [37] found that more complex AI models do not necessarily achieve higher accuracy than traditional statistical approaches. Similarly, Claveria et al. [33] reported promising results but emphasized that the predictive reliability of neural networks is highly dependent on the quality of data preprocessing. Conversely, several studies have demonstrated that AI-driven forecasting methods can outperform conventional techniques. For example, Sun et al. [38] applied machine learning to predict tourist arrivals, while Law and Au [39] and Huang [40] utilized neural networks to forecast travel demand, further reinforcing the potential of AI in tourism forecasting. Nonetheless, careful model selection and rigorous validation remain essential to ensure the reliability and precision of AI-based forecasting in the tourism sector.

b) *Machine translation*: The travel and tourism sector frequently involves interactions in multiple languages, yet language barriers remain one of the primary challenges faced by tourists, often leading to discomfort and stress during their travels [41]. These barriers frequently discourage travelers from fully engaging with local culture, resulting in a preference for familiar international franchises and well-known brands instead of exploring authentic local experiences. Since personalization plays a significant role in helping tourists discover new places [42], machine translation serves as a valuable tool for enhancing accessibility and engagement. By facilitating real-time communication, machine translation enables travelers to navigate destinations more effectively, allowing them to participate in a wider range of activities without linguistic constraints. The incorporation of artificial intelligence, particularly through machine learning and natural language processing (NLP), has greatly improved the development of machine translation applications and real-time translation systems, further enriching the tourism experience.

c) *Robots and integrated systems*: Robots, as autonomous machines equipped with artificial intelligence, are designed to perceive their surroundings, decide, and execute accordingly. A key distinction between robots and other AI-driven technologies lies in their physical embodiment, while their autonomy differentiates them from devices such as ATMs, self-service check-in points. Robots could be categorized into various types [43], including humanoid, animal-like, object-based, or functionally designed models. Historically, robots were predominantly deployed in industrial settings; however, advancements in artificial intelligence have facilitated their integration into service-oriented industries [44]. In tourism, service robots are increasingly addressing common challenges such as overcoming language barriers and mitigating labor shortages, providing innovative solutions that enhance operational efficiency and customer experiences [45].

d) *Recommendation systems and Personalization*: For a long time, tourists relied on printed directories and static images to determine which destinations to visit, as well as to select points of interest and activities available at their chosen location. The Internet has increased the amount of information available, and user-generated content (UGC) has also helped tourists make better, more informed decisions. However, even given the availability of this additional information, price was the most important factor in decision-making. Artificial intelligence is changing this behaviour as it allows tourists to find the best alternatives and allows businesses to tailor their experiences to the particular requirements of their customers. This is achieved through recommendation systems and personalization. These techniques give travellers options that best match their interests [46]. The use of recommendation systems in tourism has gained increasing importance as the abundance of choices for users has grown exponentially with online environments [47]. Recommender systems generally work by aligning the characteristics of available options with user profiles to suggest the most suitable choices. Personalization techniques aim to deliver tailored information to users based on their preferences and constraints [48]. As a result, businesses transition from broad, mass-market strategies to individualized marketing approaches. Implementing personalization requires extensive data on user behavior to accurately construct detailed user profiles.

e) *Chatbots*: Chatbots enable customers to participate in conversations primarily focused on information retrieval. These interactions can be time-consuming and may encompass multiple processes [2]. Chat systems are often known as chatbots or virtual agents [49]. These systems incorporate technologies such as natural language processing (NLP) and speech recognition, making them widely prevalent. They are commonly found in personal assistants on smartphones and smart home devices, including commercial platforms like Apple Siri, Google Assistant, Microsoft Cortana, and Amazon Alexa. Additionally, text-based chatbots are integrated into websites and kiosks, enhancing customer interactions. As these systems require minimal user effort and simulate human-like communication, they have become a standard in digital interactions. Melián-González et al. [17] analyze the key factors influencing chatbot adoption.

The future of artificial intelligence in tourism remains uncertain, with both optimistic and cautious perspectives. From an optimistic standpoint, AI advancements could address key societal challenges. Privacy concerns may be effectively managed through robust regulatory frameworks, enhanced security measures, and transparent data governance. Additionally, seamless connectivity could facilitate the development of integrated IT systems, enabling greater efficiency in tourism operations. In this scenario, human workers and AI-driven systems would collaborate harmoniously, complementing each other's strengths and optimizing service delivery. Within this framework, AI is viewed as a collection of technologies that will enhance the tourism experience, benefiting all stakeholders. Businesses will gain deeper insights into customer needs, allowing them to develop products, services, and tourism experiences that are more closely aligned with individual preferences. Additionally, AI will enable businesses to dynamically generate personalized travel packages tailored to customer interests. Technologies will both substitute and enhance various tasks, streamlining operations and lowering overall costs. These efficiency gains can translate into financial savings, some of which may be passed on to consumers. As a result, businesses will be able to provide more cost-effective services, making offerings that were once considered financially inaccessible more affordable to a broader range of travelers [45]. Technology can also enhance specific jobs or relieve employees of certain duties, improving customer experience and satisfaction. From the customer's perspective, AI will enable faster trip planning, significantly reducing transaction costs while offering highly personalized travel packages tailored to their specific preferences and interests. In addition to the duration of the trip, the use of advanced technology will decisively help tourists navigate seamlessly in unfamiliar environments, reducing anxiety and fear of the unknown. Language and cultural differences will not be obstacles to tourism; instead, they will be an additional attraction that remains to be explored. To ensure the beneficial applications of artificial intelligence in the tourism industry, research is needed to understand the effects of artificial intelligence [29, 40]. Studies should focus on the impact of AI in the tourism sector, how AI can improve customer experiences and the impact of AI on the tourism industry [28, 50]. In addition, research should explore the intersection between digital technologies and tourism experiences and how artificial intelligence can be used to create a sustainable future for the tourism industry [51].

5 The Challenges Facing the Tourism Industry in Integrating Artificial Intelligence

In recent years, it is evident that artificial intelligence has penetrated the wider tourism community, particularly in service industries such as hospitality and tourism [23]. The main research question is what challenges does the tourism industry face when incorporating artificial intelligence [29]. The tourism industry faces significant challenges when it comes to integrating artificial intelligence. These include the design of artificial intelligence systems, the difficulty in its acceptance by tourism businesses, the need to assess its effects carefully, as well as the practical application of it and related technologies in the tourism sector, such as in tourism businesses. Overall, AI can be a powerful tool that can be used to improve customer experience and operational efficiency in the tourism industry.

Using artificial intelligence in the tourism sector presents both opportunities and challenges. From the perspective of the tourism industry, the opportunities for harnessing artificial intelligence are enormous, as it can shape the future of tourism as a social phenomenon and economic activity. However, there are several challenges facing the tourism industry when integrating AI, such as designing beneficial AI, facilitating adoption, assessing the impact of AI, and implementing AI and its technologies. Furthermore, using artificial intelligence in the tourism industry is part of a more significant technological revolution that has been going on for decades. This kind of technological revolution has brought new ideas such as the "sharing economy", the "internet of things" and the "internet of people" to the tourism industry, further increasing the potential opportunities for the tourism industry to leverage the use of artificial intelligence. Research is needed to understand its impact to ensure the beneficial applications of AI in the tourism industry. The findings of this research paper highlight the need for systematic knowledge generation to ensure beneficial applications of intelligent automation in the tourism industry. Overall, AI can be a powerful tool that can be used to improve customer experience and operational efficiency in the tourism industry. However, addressing the challenges and implementing AI best to serve the industry's and its stakeholders' needs is important. Further research and collaboration between AI experts and tourism professionals is needed to achieve this goal.

One of the key challenges hotels face is that big data plays a fundamental role in the advancement of artificial intelligence. However, categorizing hotel data sets as big data is not straightforward. While hotels typically collect large volumes of data with some degree of variety, this data is often restricted to guests' interactions with the hotel website before their stay and their behavior within the hotel itself. Hotels generally lack comprehensive information about their guests' profiles, interests, preferred destinations, and behaviors outside the hotel environment. To obtain a more extensive dataset, hotels must collaborate with external businesses that can supplement their data. A significant concern in this regard is consumer privacy [52], as well as the uncertainty surrounding the rapid development and acceptance of AI [23], which may slow down its wider adoption and integration within the hotel industry.

In the final stage of AI development and full implementation within the hospitality industry, where systems are fully integrated and operational, the majority of tasks currently performed by humans will be handled by robots and AI-driven systems. However, this does not imply that the hospitality sector will operate entirely without human presence. Humans will continue to play a crucial role in two key areas. First, they will manage a select set of tasks that remain highly complex and challenging to automate, even with advancements in robotics and AI capabilities. Second, human presence will serve as a distinguishing factor and a symbol of luxury, adding value to the service experience. In cases where human labor is less economically efficient, its role will be justified either by its ability to enhance service differentiation or by contributing to higher service quality, much like the current practice in certain gas stations [17].

In conclusion, technological advancements will enable customers to receive high-quality service while striving to uphold privacy protections as much as possible [45]. However, a more pessimistic perspective on AI adoption also exists. According to this view, widespread automation will lead to the replacement of many industrial jobs, displacing workers whose roles traditionally required human involvement. Additionally, regulations may mandate the use of AI-driven machines, as already seen in numerous airports. Tourists may find themselves interacting with machines and robots that are not yet fully optimized for real-world applications, while businesses might prioritize cost-saving measures by opting for lower-quality automated solutions that fail to match the service standards of human employees. In most cases, technology will take over human roles, yet the anticipated reduction in labor costs may not necessarily result in lower prices for consumers. Instead, customers may continue to pay the same rates while experiencing a decline in service quality. Moreover, concerns about data privacy and security will persist, as AI integration may not fully safeguard sensitive customer information. Workers may struggle to adapt to AI-driven environments, and many organizations may face difficulties in effectively incorporating AI systems into their operations. As with most transformative technologies, neither the optimistic nor the pessimistic scenario is likely to unfold in its entirety. Instead, the future will likely be shaped by a combination of both perspectives, with AI presenting both opportunities and challenges for the hospitality and tourism industries.

6 Methodology

This study employed a qualitative research approach, utilizing semi-structured interviews to explore the perspectives of tourism professionals on integrating artificial intelligence (AI) in the tourism industry. Semi-structured interviews were chosen for their flexibility, allowing participants to share detailed insights while enabling the researcher to probe specific themes.

6.1 Data Collection

The study involved four tourism professionals selected based on their experience and willingness to participate voluntarily. Each participant received a detailed information form outlining the study's objectives, confidentiality measures, and their rights as participants, including the option to withdraw at any stage. Anonymity and confidentiality were ensured by assigning unique codes to participants (e.g., Participant 1) and omitting any personally identifiable information. Interviews were conducted online or face-to-face, depending on participants' preferences, and lasted an average of 30–60 min. Open-ended questions guided the discussions, focusing on themes such as AI's influence on job roles, operational efficiency, customer experiences, and ethical implications. Participants were encouraged to share their insights freely, allowing the researcher to probe further where necessary. While most interviews were recorded for transcription, participants who opted out of recording were accommodated with detailed notetaking.

6.2 Data Analysis

Thematic analysis was employed to identify, analyze, and report patterns within the data. This flexible method allowed the researcher to extract recurring themes and insights without adhering to a predetermined theoretical framework [53]. The analysis process involved transcribing recorded interviews and coding the data inductively. Themes were then categorized into six core areas: process automation, perceptions of AI, future industry direction, ethical considerations, workforce training, and strategic decision-making.

6.3 Limitations

The primary limitation of this study is the small sample size, which restricts the generalizability of the findings. The relatively limited body of literature on AI in the tourism industry also posed challenges in contextualizing the results. Despite these limitations, this research contributes valuable qualitative insights to an emerging field and provides a foundation for future studies.

7 Results and Discussion

The qualitative analysis of interviews with tourism professionals highlighted six core themes that illustrate the multifaceted impact of artificial intelligence (AI) on the tourism industry. These themes provide valuable insights into AI's benefits, challenges, and transformative potential in this sector.

7.1 Impact of AI on Work

Participants emphasized AI's significant role in automating repetitive processes, reducing the need for human intervention. This shift allows employees to focus on more creative and strategic tasks. For example, AI-driven platforms can propose automated pricing adjustments based on demand, streamlining operational decision-making. However, the introduction of AI also necessitates workforce adaptation. New skills and training are required as roles and responsibilities evolve, including data management and analysis. While some participants expressed concerns about job displacement, they also acknowledged the creation of new roles, such as AI maintenance and data analytics (Table 1).

Table 1. Impact of AI on Work.

Theme	Underlay	Quote/researcher's note
Effects of AI on Work	Automation of processes	*"AI enables the automation of many processes, reducing the need for human intervention."* (Participant 1)
	New skills and training needs Changes in roles and responsibilities	*"Employees must develop new skills to remain competitive."* (Participant 2) *"Roles are changing, with more data management and analysis tasks. Employees must be ready to take on more responsibility and adapt to new work realities."* (Participant 3)
	Job loss and job creation	*Researcher's note: Routine tasks will be removed such as check in/out at hotel units, reservation management* *"Creating new jobs such as maintenance of artificial intelligence and robotics systems for tourism." (*Participant 4*)*

Source: Authors' analysis

AI has transformed work processes in the tourism industry, with the following subthemes emerging:

- ***Process automation***: The impact of AI on work is diverse as it changes the reality of existing social work. One of the main applications of artificial intelligence is the automation of repetitive processes, where previously, they required a lot of time and human resources, which now have the ability to deal with more creative objects. Machine learning systems and algorithms, as mentioned both in theory and in the first interview, can draw important conclusions from large data sets, helping the tourism industry better understand the needs and desires of its customers. Thus, businesses have the ability to produce more specialized solutions and products that improve the customer experience as well as the entire tourism industry. Participant 1 gave a typical example, pointing out that AI enables the automation of many processes, reducing the need for human intervention.

- ***Changes in roles and responsibilities***: The integration of AI necessitates a shift in job roles, with employees required to adapt to data-driven responsibilities. Participant 3 gave a typical example, pointing out that roles are changing, with more data management and analysis tasks, and that employees must adapt to new work realities.
- ***Job loss and job creation***: While some routine jobs are being replaced, new roles in AI system maintenance and data analytics are emerging. So, AI undoubtedly has the potential to redefine roles and responsibilities in the tourism industry. Participants 2 and 4 gave typical examples, pointing out that AI will create new jobs in the technology sector, such as data analysts and algorithm specialists, but also replace other jobs with routine tasks.

Therefore, AI is transforming work processes in the tourism industry by automating repetitive tasks, enabling employees to focus on creative and strategic responsibilities. As noted in the interviews, participants highlighted the importance of reskilling the workforce to adapt to roles in data analysis and AI system maintenance. For example, Participant 1 emphasized that AI enables the automation of many processes, reducing the need for human intervention. This finding aligns with Zsarnoczky [28], who observed that AI reshapes job roles by automating routine tasks and creating new opportunities in technology-focused areas like algorithm management and data analytics. Similarly, Bowen & Morosan [45] predicted a redefinition of jobs in the hospitality sector due to AI integration, further corroborating the participants' insights.

7.2 Perceptions of AI

The professionals interviewed had a generally positive outlook on AI, citing its ability to enhance customer service, unlock innovative opportunities, and improve efficiency. However, concerns were raised regarding potential job losses and the need for ongoing training to ensure AI technologies' safe and effective use. Participants stressed the importance of balancing technological advancement with human oversight (Table 2).

Participants shared mixed perceptions about AI, categorized into the following subthemes:

- Positive perceptions: AI was viewed as a tool to enhance service quality and unlock innovative opportunities. Young professionals in the tourism industry have an overall positive assessment of artificial intelligence, as they see new opportunities for growth, improvement of services provided, and new possibilities and innovations emerge. Participants 1 and 2 noted that AI could help provide better customer service and unlock possibilities they could not imagine.
- Concerns: Participants highlighted potential job losses, data security risks, and the need for training in safe AI practices. Participant 4 noted concern about job losses due to automation.

Thus, participants expressed a generally positive perception of AI's potential to enhance customer service and unlock innovative opportunities. However, concerns about job displacement and data security persisted. For instance, Participant 4 mentioned the

Table 2. Perceptions of AI.

Theme	Underlay	Quote/researcher's note
Perceptions of AI	Positive approach	*"AI can help provide better customer service."* (Participant 2) *"It can unlock possibilities that we don't even imagine."* (Participant 1)
	Concerns	*"There is concern about job losses due to automation."* (Participant 4) *"Even if jobs are cut in some places, like in customer support, the proper management of technology will still need people."* (Participant 2) *"There is also a need for ongoing training on the safe use of AI so that professionals are aware of current practices and the technological threats they may face."* (Participant 2)

Source: Authors' analysis

potential for job losses due to automation, a concern echoed in studies by Prentice et al. [23]. The optimism around AI's ability to deliver tailored services and reduce operational costs aligns with findings by Buhalis & Cheng [15], who emphasized AI's role in improving service delivery and operational efficiency. These perspectives highlight the dual nature of AI as both an opportunity and a challenge for the industry.

7.3 Future Direction of the Tourism Industry

AI was seen as a driver of innovation, enabling personalized services and more efficient operations. Participants noted that investments in AI technologies could reduce operational costs and potentially lower the price of travel packages. They also emphasized the need for businesses to adapt to the increasing integration of AI to remain competitive and gain new, unique, comparative advantages. According to Participant 1: "AI tools can analyze huge amounts of data in real-time, which was impossible to do by humans alone. This means we can react more quickly to market changes and adjust our strategies accordingly". At the same time, many participants pointed out that the cost of tourist packages may be reduced for the consumer/tourist/customer as operational costs will be reduced due to the automation of processes and the loss of staff. Also, through the optimization of demand forecasting processes, it will save time and resources. Bibliographically, this is confirmed, as Zsarnoczky [28] has pointed out that artificial intelligence can be used to improve the efficiency of operations and reduce costs in the tourism industry (Table 3).

Key insights into AI's role in shaping the future of tourism were grouped as follows:

- ***Innovations***: AI facilitates the personalization of services and enhances operational efficiency. For example, Participant 3 stated that using AI leads to innovations such as the personalization of services.

Table 3. The future direction of the tourism industry with AI.

Theme	Underlay	Quote/researcher's note
The future of the tourism industry with the continuous use of AI	Innovations	*"The use of AI leads to new innovations such as personalization of services."* (Participant 3)
	Investments and development	*"Companies are investing more in technology to compete effectively."* (Participant 1) *"The booking engines of the big travel agents will evolve."* (Participant 2)
	Tourist package pricing	*"Although it may sound strange, artificial intelligence has the potential to reduce the cost of travel packages, rather than increase them. This is because artificial intelligence can automate many processes and improve efficiency."* (Participant 1) *"On the contrary, it can lead to reduced operating costs due to the automation of processes. AI enables time and resource savings through process optimization and demand forecasting, without necessarily increasing prices for consumers."* (Participant 2) *"Can increase the price of tourist packages through customized and autonomous service."* (Participant 4)

Source: Authors' analysis

- ***Investments and development***: Increasing investments in AI technologies enable businesses to stay competitive. For example, Participant 1 stated that companies invest more in technology to compete effectively.
- ***Tourist package pricing***: AI can reduce operational costs, potentially lowering travel package prices. For example, Participant 2 stated that AI can automate processes, reducing costs and potentially lowering consumer prices.

Consequently, AI is seen as a catalyst for innovation and operational efficiency in tourism. Participants noted its potential to lower travel package costs through process automation and demand forecasting. For example, Participant 2 explained how AI enables time and resource savings through process optimization. Literature supports these insights, with Zsarnoczky [28] highlighting AI's capacity to reduce costs and drive innovation in tourism. Additionally, studies by Gaur et al. [22] reveal how personalized AI-driven recommendations can enhance customer satisfaction and operational profitability.

7.4 Ethics and Privacy

Ethical considerations emerged as a key theme, particularly the importance of safeguarding customer privacy and ensuring the responsible use of AI. Participants advocated for clear regulations and institutional frameworks to address ethical challenges and mitigate potential risks (Table 4).

Table 4. Ethics and privacy in AI.

Theme	Underlay	Quote/researcher's note
Ethics and privacy in AI	Privacy and protection of personal data	*"We need to ensure that AI is used in ways that protect customer privacy."* (Participant 4)
	Ethical concerns	*"The development of AI must take into account ethical issues and its responsible use" "But this also depends on the correct institutionalization of its use as is done with the internet which can have both good and bad use."* (Participant 2)

Source: Authors' analysis

From the ethical considerations surrounding AI in tourism that emerged as a critical concern, the following sub-themes were identified:

- ***Privacy and protection of personal data***: Participants emphasized the importance of ensuring that AI systems prioritize customer privacy and secure data management. For example, Participant 4 stated that there is a need to ensure that AI is used in ways that protect customer privacy.
- ***Ethical Concerns***: Concerns were raised about AI's ethical implications and the need for clear regulations to guide its application in tourism. For example, Participant 2 stated that the development of AI must consider ethical issues and its responsible use.

Therefore, ethical concerns surrounding AI were a key focus, particularly regarding privacy and data security. Participant 4 emphasized that AI must prioritize customer privacy and ensure secure data management. This aligns with Tussyadiah [29], who stressed the importance of regulatory frameworks to address ethical challenges in AI implementation. Ensuring responsible use of AI is crucial for building trust and achieving long-term sustainability in the tourism industry.

7.5 Training and Workforce Adaptation

The interviews highlighted the necessity of continuous learning and workforce reskilling to prepare for an AI-driven industry. Participants suggested that businesses should invest in training programs to equip employees with the skills needed to collaborate effectively with AI systems (Table 5).

Table 5. Training and adaptation of workers.

Theme	Underlay	Quote/researcher's note
Training and adaptation of employees	Training programs from companies in the sector	*"Businesses must invest in training programs."* (Participant 1)
	Lifelong learning	*"Continuous learning is key to success in the age of AI."* (Participant 3)

Source: Authors' analysis

The need for employee training and adaptation in an AI-driven industry was a recurring theme, divided into the following sub-themes:

- ***Training programs from tourism companies***: Participants highlighted the importance of industry-led initiatives to upskill employees in AI technologies. After artificial intelligence enters the tourism sector, workers will have to be retrained and adapt to the new working reality. For example, Participant 1 noted that businesses must invest in training programs.
- ***Lifelong learning***: Continuous education was identified as essential for staying competitive in a rapidly evolving technological landscape. In today's era, where AI has entered the tourism industry very strongly, employees' lifelong learning should be at the centre of young people's efforts to stay competitive and develop their skills in tourism. Characteristically, Participant 1 noted a constant need for training and skill development, while Participant 3 stated that continuous learning is key to success in the age of AI.

Thus, continuous learning and workforce reskilling were highlighted as essential for adapting to AI-driven changes. Participants emphasized the need for industry-led training initiatives to upskill employees. These findings are consistent with Huang et al. [40], who state the need for a robust framework to equip tourism professionals with the necessary skills to leverage AI technologies effectively.

7.6 Strategy and Decision-Making

AI's ability to process and analyze large volumes of data was identified as a significant advantage for strategic planning. By leveraging AI tools, businesses can make more accurate, data-driven decisions, anticipate market trends, and improve competitiveness (Table 6).

Table 6. Strategy and decision-making.

Theme	Underlay	Quote/researcher's note
Strategy and decision-making	Improvement of decision-making Analysis of larger volumes of data	*"AI helps make better data-driven decisions."* (Participant 2) *"The ability to analyze large volumes of data is an important factor."* (Participant 4)

Source: Authors' analysis

The analysis of AI's role in improving strategic decision-making in tourism businesses concluded with the following sub-themes:

- ***Improvement in decision-making***: AI has proven to be an important ally in tourism businesses' decision-making process. By efficiently processing large volumes of data, AI facilitates more accurate and informed decisions. It can also quickly and efficiently process all the information, allowing officials to distinguish upcoming trends from trends that are not immediately apparent. For example, Participant 2 noted that AI helps make better data-driven decisions.
- ***Analysis of larger volumes of data***: The fact that it is now possible to process larger volumes of data in a shorter period of time is a catalytic factor for both strategy and decision-making. The ability to analyze extensive datasets enables businesses to identify trends and adapt their strategies accordingly. The analysis performed on the data collected from accommodation reservations and personal searches of customers or comments they may make (e.g., Google reviews) provides useful information regarding customer preferences and upcoming trends in this sector's market. Thus, AI enables industry professionals to adapt their services to improve competitiveness continuously. For example, Participant 4 noted that the ability to analyze large volumes of data is an important factor.

So, AI's ability to process large volumes of data enables businesses to make informed, data-driven decisions. Participant 2 noted that AI helps make better data-driven decisions, highlighting its role in strategic planning. Research by Wirtz [25] corroborates this, demonstrating how AI supports improved decision-making and trend analysis in tourism. By leveraging AI tools, businesses can enhance competitiveness and adapt to rapidly changing market dynamics.

7.7 Implications of the Results

The findings of this study reveal that AI has the potential to revolutionize the tourism industry, creating both opportunities and challenges for industry stakeholders. AI's ability to automate repetitive tasks and analyze vast datasets enables businesses to streamline operations and enhance customer experiences. Participants emphasized AI-driven innovations such as predictive analytics, chatbots, and recommendation systems. These technologies, as corroborated by Buhalis & Cheng [15], allow businesses to offer tailored travel solutions while reducing operational costs. For example, AI-powered systems like those implemented by British Airways' "Know Me" program improve customer engagement through personalized interactions. This shift calls for businesses to invest in AI infrastructure and training programs. Hotels, airlines, and travel agencies can use AI to optimize resource allocation, such as adjusting pricing dynamically based on demand forecasting [25]. Personalization techniques, like those used by AI-based recommendation systems in the hotel industry [22], can significantly improve customer satisfaction and loyalty. Moreover, AI's integration into the tourism industry necessitates workforce adaptation. Participants noted that while routine jobs may be displaced, new roles in AI maintenance, data analytics, and technology management are emerging. This aligns with Bowen & Morosan's [45] observation that AI will redefine roles in the hospitality sector. Businesses must, therefore, prioritize continuous training and upskilling programs to equip employees for an AI-driven future. Furthermore, ethical concerns surrounding AI, particularly regarding privacy and data security, are critical for the tourism industry. As Participant 4 emphasized, safeguarding customer privacy is paramount. Tussyadiah [29] also highlights the need for robust regulations to ensure responsible AI use. Businesses must implement clear data protection policies and communicate these measures to customers to build trust.

The study underscores the need for comprehensive policymaking to guide AI's ethical and practical implementation in tourism. The ethical considerations highlighted in the study, such as data privacy and responsible use, require policymakers to establish clear guidelines for AI adoption. Governments should collaborate with industry stakeholders to develop frameworks that address these issues. For example, Europe's General Data Protection Regulation (GDPR) provides a benchmark for managing data privacy in AI systems. Governments can play a pivotal role in accelerating AI adoption by offering financial incentives, such as tax breaks or grants, for businesses investing in AI technologies. These incentives can help small and medium-sized enterprises (SMEs) overcome initial barriers to adopting advanced AI systems. Moreover, policymakers should support workforce transformation through funding for vocational training and AI literacy programs. Establishing partnerships between educational institutions and industry leaders can ensure that training programs are aligned with industry needs. This approach aligns with Huang et al.'s [40] framework for preparing professionals for AI-driven industries.

The findings of this study also open several avenues for future research. Further research is needed to understand customer perceptions of AI-driven services in the tourism industry. Studies could explore how customers evaluate AI-based personalization compared to traditional human interactions and its impact on satisfaction and loyalty. Moreover, the workforce transformation identified in this study highlights the need for longitudinal research to assess the long-term effects of AI on employment in tourism. Future studies could examine how job roles evolve and whether the benefits of AI outweigh its disruptive effects. Additionally, sustainability is a growing concern in the tourism industry. Research should investigate how AI can contribute to sustainable tourism practices, such as optimizing resource use, reducing waste, and promoting eco-friendly travel options.

Finally, industry stakeholders should prioritize investments in AI infrastructure to enhance operational efficiency and improve customer satisfaction. Equally important is implementing continuous training and reskilling initiatives, equipping employees with the skills required for AI-integrated roles. A strong focus on ethics and privacy is essential, with businesses adopting robust data protection policies and transparently communicating these measures to build customer trust. Collaboration with policymakers is also critical; industry leaders must work alongside governments to develop regulatory frameworks and workforce development initiatives that ensure the responsible and effective integration of AI technologies in the tourism sector.

8 Conclusions

This study explored the transformative impact of artificial intelligence (AI) on the tourism industry, highlighting its potential to enhance customer experiences, streamline operations, and foster innovation. Through a qualitative approach involving semi-structured interviews with tourism professionals, six key themes were identified: the impact of AI on work, perceptions of AI, future industry directions, ethical considerations, workforce training, and strategic decision-making.

The findings confirmed that artificial intelligence can significantly improve decision-making processes. It allows companies to adjust their services and strategies using more accurate and meaningful data, thus strengthening their comparative advantage. The findings also demonstrated that AI-driven technologies significantly improve operational efficiency by automating routine tasks and enabling businesses to provide highly personalized services. For example, AI-powered chatbots and recommendation systems streamline customer interactions while predictive analytics optimize resource allocation and pricing strategies. These advancements align with existing literature, such as Buhalis & Cheng [15] and Wirtz [25], which underscore AI's role in driving operational and customer-centric innovation.

However, the integration of AI presents challenges, including workforce adaptation and ethical considerations. As roles evolve to include data analytics and AI system management, there is a critical need for continuous training and workforce reskilling. Based on the findings, the effects of AI at work concern the automation of processes, the new skills and training needs that are born, and the changes in roles and responsibilities. The participants also mentioned the loss of certain positions as well as the creation of new jobs. Research by Prentice [23] and Tussyadiah [29] confirms the trend towards automation of some "routine" tasks, as found in the research findings. As early as 2018, Bowen & Morosan [45] pointed out that this new technology, as it is integrated into the tourism industry, will redefine jobs and create new ones, which the participants of the present also confirm. Nevertheless, the new educational needs and the cultivation of the skills required are not mentioned anywhere in the literature. They complement the existing research on using artificial intelligence in the tourism industry. In this way, the need for continuous training and preparation of professionals in the sector becomes clear in order to effectively utilize the new possibilities offered by artificial intelligence while, at the same time, it is also evolving.

In relation to the future of the tourism industry, the participants focused on the innovations it brings, the upcoming investments as well as the changes in the price of the tourist package, which is also confirmed in the literature by Zsarnoczky [28], according to which the reduction of costs in the tourism industry will also bring about a reduction in the price of the tourist package. Additionally, concerns about data privacy and the ethical use of AI highlight the importance of regulatory frameworks to ensure responsible implementation. Further focus should be placed on safeguarding privacy and personal data as it emerges as a critical issue, requiring comprehensive regulations (legislation) and responsible use of technology. In addition, the ethical considerations surrounding the use of artificial intelligence require the utmost attention, ensuring that its advancement and integration adhere to responsible and ethical practices. These findings echo the work of Tussyadiah [29], who emphasized the need for ethical AI practices in tourism.

Technological development, especially artificial intelligence, has dynamically entered the Greek tourism industry, offering significant opportunities for growth and progress. With its appropriate institutionalization and responsible use, in the future, it can improve the customer experience, strengthen the competitiveness of businesses, and offer new, innovative tools to improve the services provided to tourists.

Policymakers and industry stakeholders must collaborate to address these challenges. Governments can play a pivotal role by incentivizing AI adoption, funding workforce training programs, and establishing robust regulations to protect customer privacy and foster trust. Industry leaders should focus on investing in AI infrastructure, adopting sustainable practices, and leveraging AI to create personalized and eco-friendly travel experiences.

Future research should delve deeper into AI's long-term impacts on employment, customer acceptance of AI-driven services, and its role in sustainable tourism and crisis management. Understanding these dimensions will provide a more comprehensive framework for integrating AI into the tourism industry.

In conclusion, AI is poised to redefine the tourism sector by enhancing efficiency, driving innovation, and creating personalized experiences. However, its transformative potential can only be fully realized through ethical implementation, workforce adaptation, and stakeholder collaborative efforts. The tourism industry can harness AI to meet travellers' evolving demands by addressing these priorities while ensuring sustainability and inclusivity.

Appendix: Interview Guide

1. How do you see artificial intelligence changing the roles and responsibilities of jobs in the tourism industry?
2. How has artificial intelligence contributed to speeding up and simplifying business processes?
3. What gaps does it come to cover in order to give surplus value to decision-making and the execution of specific procedures?
4. Since when has artificial intelligence been used in the tourism industry?
5. What are the opportunities and threats from what use of it?
6. Has the use of artificial intelligence led/will it lead to an increase in the price of the tourist package?
7. In which types of products and services can we distinguish the use of AI in tourism? Has its use been safely evaluated to draw a safe conclusion about its benefits?

8. What threats does the traditional work model (in tourism) face, and what would its use mean for jobs in a tourism business?
9. How safe can an artificial intelligence system be at the level of customer information and personal data?
10. In which areas is there a clear improvement from a technological point of view, and what possibilities will artificial intelligence provide to customers and businesses in the near future?
11. What could be done so that jobs remain unaffected and harmonious cooperation between human and technological factors continues?
12. Artificial intelligence could contribute to the future forecast of demand (for a service, a product, a destination, etc.).
13. Consequently, it could predict the threats and, therefore, the weak points that need to invest resources (financial, human, etc.) either in a destination as a whole or in a tourism product
14. What occupations/types of jobs will the widespread adoption of AI by all tourism businesses eliminate and create?

References

1. Rudas, I.J., Fodor, J.: Intelligent systems. Int. J. Comput. Commun. Control III (Suppl.). Proceedings of ICCCC 2008, pp. 132–138 (2008)
2. Gretzel, U.: Intelligent systems in tourism: a social science perspective. Ann. Tour. Res. **38**(3), 757–779 (2011)
3. Chen, Y.-C., Ahn, M.J., Wang, Y.-F.: Artificial Intelligence and public values: value impacts and governance in the public sector. Sustainability **15**(6), 4796 (2023)
4. Brooks, R.: Artificial intelligence is a tool, not a threat (2014). https://robohub.org/artificial-intelligence-is-a-tool-not-a-threat/. Accessed 12 Nov 2024
5. Poole, D.L., Mackworth, A.K.: Artificial Intelligence: Foundations of Computational Agents, 2nd edn. Cambridge University Press (2017)
6. Rane, N.L., Mallick, S.K., Kaya, O., Rane, J.: Applications of machine learning in healthcare, finance, agriculture, retail, manufacturing, energy, and transportation: a review. In: Applied Machine Learning and Deep Learning: Architectures and Techniques, pp. 112–131. Deep Science Publishing (2024)
7. Sterne, J.: Artificial Intelligence for Marketing: Practical Applications. Wiley, Hoboken (2017)
8. Millauer, T., Vellekoop, M.: Artificial intelligence in today's hotel revenue management: is it worth noting. Res. Hospitality Manag. **9**(2), 121–124 (2019)
9. Viverit, L., Heo, C.Y., Nobre Pereira, L., Tiana, G.: Application of machine learning to cluster hotel booking curves for hotel demand forecasting. Int. J. Hosp. Manag. **111**, 103455 (2023)
10. Wooldridge, M.: The Road to Conscious Machines: The Story of AI. Penguin (2020)
11. Berrar, D.: Machine Learning with R: Expert Techniques for Predictive Modeling, 3rd edn. Packt Publishing (2021)
12. Alpaydin, E.: Machine Learning. MIT Press (2016)
13. Mastrogeorgiou, G.: Analysis: what is Artificial Intelligence and how will it change our lives? (2018). https://todiktio.eu/analysi-ti-einai-i-texniti-noimosyni-kai-pos-tha-allaksei-ti-zoi-mas/. Accessed 20 Nov 2024

14. Dwivedi, Y.K., et al.: Artificial Intelligence (AI): multidisciplinary perspectives on emerging challenges, opportunities, and agenda for research, practice and policy. Int. J. Inf. Manage. **57**, 101994 (2021)
15. Buhalis, D., Cheng, E.S.Y.: Exploring the use of chatbots in hotels: technology providers' perspective. In: Information and Communication Technologies in Tourism 2020: Proceedings of the International Conference in Surrey, United Kingdom, 08–10 January 2020, pp. 231–242. Springer, Cham (2020)
16. Samara, D., Magnisalis, I., Peristeras, V.: Artificial intelligence and big data in tourism: a systematic literature review. J. Hosp. Tour. Technol. **11**(2), 343–367 (2020)
17. Melián-González, S., Gutiérrez-Taño, D., Bulchand-Gidumal, J.: Predicting the intentions of using chatbots for travel and tourism. Curr. Issue Tour. **24**(2), 192–210 (2019)
18. Yun, J., Park, J.: The effects of chatbot service recovery with emotion words on customer satisfaction, repurchase intention, and positive word-of-mouth. Front. Psychol. **13**, 922503 (2022)
19. Pereira, T., Limberger, P.F., Minasi, S.M., Buhalis, D.: New insights into consumer intention. To continue using chatbots in the context of tourism. J. Qual. Assur. Hospitality Tourism **25**(4), 754–780 (2022)
20. Li, J.J., Bonn, M.A., Ye, B.H.: Hotel employee's artificial intelligence and robotics awareness and its impact on turnover intention: the moderating roles of perceived organizational support and competitive psychological climate. Tour. Manag. **73**, 172–181 (2019)
21. INSETE: Action Plans to strengthen the competitiveness and structural adjustment of the tourism sector. Institute of the Association of Greek Tourist Companies (2021). https://insete.gr/wp-content/uploads/2021/12/Megatrends.pdf. Accessed 14 Nov 2024
22. Gaur, L., Afaq, A., Singh, G., Dwivedi, Y.K.: Role of artificial intelligence and robotics to foster the touchless travel during a pandemic: a review and research agenda. Int. J. Contemp. Hosp. Manag. **33**(11), 4079–4098 (2021)
23. Prentice, C., Dominique Lopes, S., Wang, X.: The impact of artificial intelligence and employee service quality on customer satisfaction and loyalty. J. Hosp. Market. Manag. **29**(7), 739–756 (2020)
24. Ashfaq, M., Yun, J., Yu, S., Loureiro, S.M.C.: I, Chatbot: modeling the determinants of users' satisfaction and continuance intention of AI-powered service agents. Telematics Inform. **54**, 101473 (2020). https://doi.org/10.1016/j.tele.2020.101473
25. Wirtz, J.: Organizational ambidexterity: cost-effective service excellence, service robots, and artificial intelligence. Organ. Dyn. **49**(3), 100719 (2019)
26. Gallego, I., Font, X.: Changes in air passenger demand as a result of the COVID-19 crisis: using Big Data to inform tourism policy. J. Sustain. Tour. **29**(9), 1470–1489 (2021)
27. Tissera, P.K., Waduge, K.T., Perera, M., Nawinna, D.P., Kasthurirathna, D.: Predictive analytics platform for airline industry. In: Proceedings of the 2020 2nd International Conference on Advances in Computing, Malabe, December 2020, pp. 108–113. IEEE, New York (2020)
28. Zsarnoczky, M.: How does artificial intelligence affect the tourism industry? Vadyba J. Manag. **31**(2), 85–90 (2017)
29. Tussyadiah, I.: A review of research into automation in tourism: launching the annals of tourism research curated collection on artificial intelligence and robotics in tourism. Ann. Tourism Res. **81** (2020)
30. Pinheiro, A., Pinto, A., Abreu, A., Costa, E., Borges, I.: The impact of artificial intelligence on the tourism industry: a systematic review. In: International Conference on Tourism, Technology and Systems, pp. 458–469. Springer, Singapore (2021)
31. Neuhofer, B., Magnus, B., Celuch, K.: The impact of artificial intelligence on event experiences: a scenario technique approach. Electron. Mark. **31**, 601–617 (2021)

32. Tuo, Y., Ning, L., Zhu, A.: How artificial intelligence will change the future of tourism industry: the practice in China. In: Wörndl, W., Koo, C., Stienmetz, J.L. (eds.) Information and Communication Technologies in Tourism 2021, pp. 83–94. Springer, Cham (2021)
33. Claveria, O., Monte, E., Torra, S.: Tourism demand forecasting with neural networks models: different ways of treating information. Int. J. Tour. Res. **17**(5), 492–500 (2015)
34. Gunter, U., Önder, I.: Forecasting international city tourism demand for Paris: accuracy of uni- and multivariate models employing monthly data. Tour. Manag. **46**, 123–135 (2015)
35. Buhalis, D., Leung, R.: Smart hospitality-Interconnectivity and interoperability towards an ecosystem. Int. J. Hosp. Manag. **71**, 41–50 (2018)
36. Stalidis, G., Karapistolis, D.: Tourist destination marketing supported by electronic capitalization of knowledge. Procedia Soc. Behav. Sci. **148**, 110–118 (2014)
37. Yu, G., Schwartz, Z.: Forecasting short time-series tourism demand with artificial intelligence models. J. Travel Res. **45**(2), 194–203 (2006)
38. Sun, S., Wei, Y., Tsui, K.L., Wang, S.: Forecasting tourist arrivals with machine learning and internet search index. Tour. Manag. **70**, 1–10 (2019)
39. Law, R., Au, N.: A neural network model for predicting Japanese demand for travel to Hong Kong. Tour. Manag. **20**(1), 89–97 (1999)
40. Huang, A., Chao, Y., de la Mora Velasco, E., Bilgihan, A., Wei, W.: When artificial intelligence meets the hospitality and tourism industry: an assessment framework to inform theory and management. J. Hospitality Tourism Insights **5**(5), 1080–1100 (2022)
41. Cohen, E.: Contemporary Tourism: Diversity and Change. Elsevier (2004)
42. Benckendorff, P.J., Xiang, Z., Sheldon, P.J.: Tourism Information Technology, 3rd edn. Cabi (2019)
43. Tung, V.W.S., Law, R.: The potential for tourism and hospitality experience research in human-robot interactions. Int. J. Contemp. Hosp. Manag. **29**(10), 2498–2513 (2017)
44. Ivanov, S.H., Webster, C.: Adoption of robots, artificial intelligence and service automation by travel, tourism and hospitality companies. In: Intelligence and Service Automation by Travel, Tourism and Hospitality Companies. Intergovernmental Scientific Conference "Modern Tourism and Service Automation and Tourism Services and Hospitality Companies-A Cost-Benefit Analysis". "Modern Tourism - Traditions and Innovations", October 2017, Sofia University, pp. 19–21 (2017)
45. Bowen, J., Morosan, C.: Attention to the hospitality industry: the robots are coming. Worldwide Hospitality Tourism Themes **10**(6), 726–733 (2018)
46. Ricci, F., Rokach, L., Shapira, B., Kantor, P.B., Shapira, B.: Introduction to recommender systems. Recommender Syst. Handb. 1–35 (2015)
47. Gavalas, D., Konstantopoulos, C., Mastakas, K., Pantziou, G.: Mobile recommender systems in tourism. J. Netw. Comput. Appl. **39**, 319–333 (2014)
48. Gao, M., Liu, K., Wu, Z.: Personalisation in web computing and informatics: theories, techniques, applications, and future research. Inf. Syst. Front. **12**(5), 607–629 (2010)
49. Buhalis, D., Harwood, T., Bogicevic, V., Viglia, G., Beldona, S., Hofacker, C.: Technological disruptions in services: lessons from tourism and hospitality. J. Serv. Manag. **30**(4), 484–506 (2019)
50. Kong, H., Wang, K., Qiu, X., Cheung, C., Bu, N.: 30 years of artificial intelligence (AI) research relating to the hospitality and tourism industry. Int. J. Contemp. Hosp. Manag. **35**(6), 2157–2177 (2023)
51. Kirtil, I., Aşkun, V.: Artificial intelligence in tourism: a review and bibliometrics research. Adv. Hospitality Tourism Rese. (AHTR) **9**(1), 205–233 (2021)
52. Hu, Y., Min, H.: The dark side of artificial intelligence in service: the "watching-eye" effect and privacy concerns. Int. J. Hosp. Manag.Manag. **110**, 103437 (2023)
53. Braun, V., Clarke, V.: Using thematic analysis in psychology. Qual. Res. Psychol. **3**(2), 77–101 (2006)

Digital Advertising and Marketing Strategies for SMEs

Benefits and Perils of SMEs' Digital Transformation in Marketing

Meletios Niros(✉), Eirini Peitzika, Angelica Niros, and Panagiota Kelali

Athens University of Economics and Business, Business Administration, 10434 Athens, Greece
mniros@aueb.gr

Abstract. It is widely accepted that digital transformation enables businesses to develop the capabilities necessary to turn weaknesses into strengths. These capabilities are required to take advantage of market opportunities and attenuate the harmful effects of threats, especially in times of turbulence. However, most of the research in digital transformation does not apply to smaller firms and the marketing capabilities created during the ongoing era of the information revolution. This chapter presents new technologies incorporated into SMEs and how they optimise marketing practices and strategies.

However, several perils may threaten SMEs' efforts to adopt digital technology in marketing. This study identifies potential perils from a consumer perspective and comprehensively analyses different aspects of consumer skepticism toward SMEs' digital transformation in marketing. This can assist future research in studying these challenges in a holistic perspective. This study also presents solutions for SMEs to overcome consumers' skepticism toward SMEs' digital transformation, which will help lead them through successful DT.

1 Introduction

In recent years, firms with less than 100 employees have not only been the backbone of both the economy and the local societies worldwide but also a significant force in the global economy (World Economic Forum 2022b). The latter businesses, also called Small and medium-sized enterprises (SMEs), represent more than 89% of the total entities globally (World Economic Forum 2022b). They contribute to about half of the total employment and 41% of total production (in value) for emergent markets (World Economic Forum 2022b). To retain growth and sustainability, SMEs must update their capabilities via digital transformation and adopt new technologies (Teng and Yang 2022). However, in an era of turbulency characterised by a great deal of volatility, uncertainty, complexity and ambiguity, the survival of SMEs has become a real challenge considering a plethora of barriers in funding and access to new technologies, skilled human resources and know-how (World Economic Forum 2022a).

Marketing, the function that is vital to produce, deliver and communicate value to customers and society in a profitable manner, is a cornerstone of the economic and social sustainability of SMEs (Ziółkowska 2021). Digital transformation, which involves integrating new information and communication technologies (NICT), new processes, and the participation of people (i.e. customers and personnel), is necessary to ensure this

A. K. Zarkada (Ed.): GMA-GAMMA 2025, SPBE, pp. 181–204, 2025.
https://doi.org/10.1007/978-3-031-96930-0_10

sustainability (Samanta et al. 2020). The process involves recording new technologies, investigating their effects on marketing practices, strategy, and performance, and understanding how customers perceive digital transformation when interacting with a small brand (Samanta et al. 2020).

In this chapter, we answer the following questions:

1. What digital technologies can SMEs use to initiate the digital transformation process in marketing?
2. What benefits can SMEs gain from incorporating Digital Transformation in their marketing tactics and strategies?
3. What are the perils of Digital Transformation in Marketing from a Consumer Perspective for SMEs?
4. What actions should be taken to alleviate these perils?

In the next section, we illustrate the digital technologies that SMEs can integrate into the digital transformation process in marketing. In the following sections, we explore the benefits and perils of this implementation from a consumer perspective and how SMEs can tackle them.

2 Digital Technologies that SMEs Can Use to Initiate the Digital Transformation Process in Marketing

Digital transformation has become vital for SMEs to adapt to rapid political, economic, societal, and technological changes and retain or increase their competitiveness (Matt and Rauch 2020). New applications and technologies appear faster than ever, enabling SMEs to apply marketing efforts more effectively and efficiently (Melo et al. 2023). This results in reaching a wider audience through digital channels and communicating brand value to customers more easily (Miklosik and Evans 2020). Digital transformation also facilitates effective feedback to SMEs, vital for continuously improving products and services (Samanta et al. 2020).

In this paragraph, we unveil the primary digital tools that speed up the process of digital transformation in marketing, such as a) *Artificial Intelligence* (AI), b) *Machine Learning* (ML), c) *Big Data Analytics*, and d) *Virtual Technology*.

The basic concept of AI is to make machines think like humans to proceed with various tasks, including decision-making and other actions requiring cognition (i.e., learning, perception, problem-solving, and understanding of language), as Verma et al. (2021) stressed. According to Cioppi et al. (2023) AI enables SMEs to apply marketing practice better and achieve their goals. It allows them to boost effectiveness and minimise costs through automation in every aspect of customer journeys enhancing customer experience as well as data-driven decision- making (Cioppi et al. 2023). AI applications on chatbots, virtual assistants, and data modeling to predict trends facilitate businesses' interaction with customers in one-to-one patterns to assist them with their purchase problems or proceed with personalised recommendations (Kotler et al. 2017). In addition, AI-powered tools optimise SMEs' marketing efforts and campaigns to achieve higher customer satisfaction and brand awareness (Huang and Rust 2021). Therefore, it can also be expected that by optimising AI, smaller businesses will be able to reach out to the desired audiences and provide and deliver higher value to the customers. The digital transformation process associated with Marketing can also be enhanced with Machine Learning, which is regarded as a subset of AI and results in newer algorithms (Zaki et al. 2023). These processes allow computers to learn based on data to improve decision-making. Supervised, unsupervised, and reinforcement learning are the primary ways to create models through ML (Zaki et al. 2023). This sophisticated technology enables even smaller firms to consider large data sets in real-time and unveil patterns of behaviours, trends, and insights required to predict customer responses related to marketing strategies and practices (Ziółkowska 2021).

Hence, the application of ML in the context of SMEs is beneficial in bringing new capabilities of analysing daily customer datasets vital to tracking new opportunity windows concerning market segments and their respective criteria about how they decide during the buying decision-making process and facilitating targeting (Miklosik and Evans 2020). According to Melo et al. (2023), ML further helps to achieve one-to-one marketing via message personalisation and case-sensitive campaigns to optimise conversion rates and marketing effectiveness (i.e. Increased Sales, Market Share and Return On Investment).

As far as big data analytics are concerned, they pose a prominent technology for SMEs regarding marketing effectiveness and efficiency (Teng and Yang 2022). The critical role of big data analytics is to rapidly transform massive amounts of data into information ready for decision-making in marketing by unveiling patterns, trend correlations, and insights (Miklosik and Evans 2020). These data sets derive from various sources, including customer interactions, website behaviour, and social media metrics. Valuable insights stem from sophisticated analytic tools and processes. The latter enables SMEs to make informed decisions about market trends and consumer behaviour and assess the outcome of their marketing campaigns in real-time (Miklosik and Evans 2020). According to Niros and Niros (2023), digitised SMEs develop a sustainable competitive advantage because they make fast and accurate decisions on marketing strategies and practices in a VUCA environment.

Further digital transformation technologies generating capabilities for marketing consist of the Internet of Things (IoT), cloud computing, blockchain technology, and Virtual Technology. IoT suggests a network of devices embedded with sensors and other sophisticated technologies, allowing them to exchange data over the internet with different systems and devices (Cioppi et al. 2023). Wearable technology and smart sensors suggest examples of IoT devices that provide SMEs with valuable data on preferences, interactions and other aspects of behaviour (Matt and Rauch 2020). Thus, IoT devices can generate new business areas for SMEs (i.e., new products) and data required to create customised and well- targeted marketing campaigns. On the other hand, cloud computing concerns creating and distributing computer services over the internet (cloud) by a cloud service provider Cioppi et al. 2023). Because of scalability and low cost, efficient launch and escalation of marketing campaigns suggest some of the most prominent benefits of cloud computing for SMEs (Matt and Rauch 2020).

On the other hand, blockchain technology has a decentralised nature. At the same time, centralised virtualisation makes a difference because it is a resulting data platform that enables the saving and retrieving of information (Matt and Rauch 2020).

Hence, blockchain technology enhances the security, transparency, privacy, and traceability required in marketing transactions to forge trust and better relationships with customers and partners. Virtual technology helps generate artificial environments, including interacting simulations and non-physically existing spaces where humans experience a situation (Cioppi et al. 2023). Augmented reality, virtual simulations, and virtual reality are some features of virtual technology that have become particularly important for marketing since they cost-effectively improve brand experience and engagement (Cioppi et al. 2023). Figure 1 below illustrates the basic digital technologies that can be used to trigger DT in marketing.

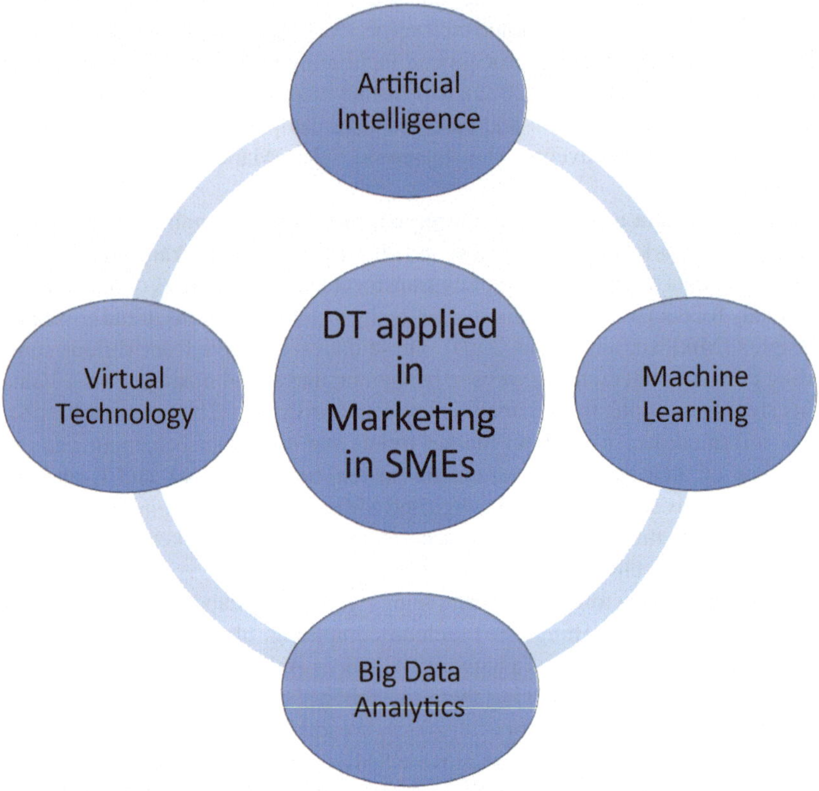

Fig. 1. Digital Technologies that SMEs can use to initiate the digital transformation process in marketing

Summarising, the embracement of digital transformation concerns a basic tool for SMEs to achieve sustainable competitive advantage and compete in new markets and pockets of growth in today's turbulent environment. In the next section, the benefits of digital transformation in terms of marketing strategies and tactics are illustrated.

3 Benefits from Incorporating Digital Transformation in SMEs' Marketing Tactics and Strategies

In the last twenty years, the marketing and business environment has become increasingly turbulent. To define a tumultuous environment, we have adopted the traits of volatile, uncertain, complex, and ambiguous (VUCA) situations that SMEs go through in terms of many difficult situations that shape their world (Zarkada 2023). Volatility refers to rapid and unpredictable situations posing a threat to organisational performance (Niros and Niros 2023). Uncertainty on the other hand, is related to the lack of information and predictability of the business environment and outcomes of actions taken by the firms (Melanthiou and Vasic, 2024). Complexity deals with a multitude of factors, often related among them, making decision-making even harder (Niros and Niros 2023). Ambiguity, which is the last element of turbulency, concerns the difficulty of fully understanding the cause and effect of situations and actions (Zarkada 2023). These external forces in a turbulent marketing environment require agile organisations to adapt quickly to change and develop an organisational culture of innovative thinking to tackle and, even better, take advantage of these external forces (Niros and Niros 2023). Digital transformation, as mentioned in the previous part, facilitates developing an *agile organisation* as soon as SMEs can become data-driven and gain access to IT infrastructure, among other capabilities (i.e. through cloud computing). Thus, they can respond rapidly to a VUCA environment and maintain their competitiveness by applying digital transformation in marketing (Melanthiou and Vasic 2024).

Since digital transformation refers to the integration of new technologies throughout the entire functions of a business, it develops new aspects in terms of operation, *value (co-) creation*, and *value delivery* to customers (Gillpatrick 2019). Increased efficiency and productivity suggest a crucial benefit of digital transformation because it optimises time-consuming tasks and processes and improves communication lines between back-office and front-line employees (De Bruyn et al. 2020). Thus, digital transformation allows SMEs and larger entities to minimise the waste of resources regarding what is required to produce and deliver a product as customers desire, increase marketing efficiency and tackle the challenges in today's global and turbulent market (Tzoumaka et al. 2024).

Concerning marketing strategies, digital transformation allows businesses to concentrate their resources on more significant and precise audiences, increasing cost-efficiency for SMEs (Niros and Niros 2023). In this sense, a high level of digitisation allows SMEs to increase customer engagement cost-effectively at all stages of the buyer decision process, thus making brand equity generation efficient (Melo et al. 2023). Developing a data- connected organisation is advantageous and can only be brought about through digital transformation. Data-driven businesses are skillful in gathering and analysing data effectively and efficiently to base their decisions on real-time information and improve their marketing strategies (Melo et al. 2023). Insights derived from digital platforms such as applications, websites, and social media can be effectively generated by (big) data analytics so as for SMEs to track consumer behaviour, desires, and aspirations and, thus, customise marketing efforts to attract and meet the demands of their target audiences (Samanta et al. 2020). Value aspects of a brand are easier to communicate to targeted segments, increasing effectiveness through the development of more accurate campaigns, and facts that increase rates of conversion and brand purchase (Niros and Niros 2023).

Marketing mix, which is particularly important in marketing, deals with marketing actions and tactics that marketing practitioners develop to differentiate their value offerings and *apply their marketing strategies more effectively* (De Bruyn et al. 2020). Marketing mix elements where actions are taken derive from the prominent 4ps model applied in products and the augmented 7ps applied in services (Niros and Niros 2023). According to the core model of 4ps, entrepreneurs need to effectively coordinate product, price, promotion, and place to forge SMEs' competitive advantage regarding market share, customer satisfaction, and loyalty (Gobble 2018). Digital transformation is once vital to gaining the *best outcome in terms of marketing mix coordination* (Gobble 2018). For example, digitisation enables SMEs to design and market new value offerings (i.e. brands) faster than ever, to proceed on the stages of new product development of product testing and iteration more rapidly based on customer perceptions and attitudes throughout a feedback process as well as to make the right decisions on pricing strategies focused on real-time competitor analysis and updated data (Rathore 2019).

Regarding promotion, digital presence in marketing platforms allows SMEs to customise promotion according to the characteristics of targeted audiences, develop campaigns, and measure everyday performance to improve future campaigns and strategies required to achieve ROI and market share goals (Rathore 2019). Improved brand visibility and experience reinforced by virtual technology, traffic diversion to digital platforms, and better conversion of leads into sales can be done more effectively for SMEs (Gillpatrick 2019). Regarding place, new technologies help SMEs maximise their digital presence in e- commerce platforms, bolster customer experience in all touchpoints, and unify customer perceptions more favourably (Niros and Niros 2023). Hence, SMEs may unify the experience provided in the entire sales channels, enabling them to divert traffic to traditional channels and sustain a unique sales proposition, posing a sustainable competitive advantage in marketing.

Similarly to marketing mix actions, smaller and medium-sized firms can implement *effective marketing strategies* due to digital transformation in marketing function (Melanthiou and Vasic 2024). Market segmentation, targeting, and Positioning have become a real-time and easy-to-analyse process since big data analytics unveil patterns and innovative segmenting variables (Kotler et al. 2017). These variables result in different purchase criteria and distinct approachability in the marketing mix previously analysed. Targeting and Positioning, which are the next steps of segmenting refer to the selection of market segment as well as the Unique Sales Proposition that is the basic reason for customers to prefer the brand compared to other competitive value offerings (Kotler et al. 2017). Digital transformation of marketing facilitates the process of targeting for exploration of potentiality of various segments become easier to occur, reducing the uncertainty of decision-making (Kotler et al. 2017).

Moreover, positioning can be used based on differentiation strategies in marketing. (Kotler et al. 2017). There are three basic positioning strategies in marketing, also known as customer equity drivers, that digital transformation can generate to boost brand preference: value equity, brand equity, and relationship equity (Niros and Niros 2023). Value Equity is related to value for money strategy, which firms implement in cases where customers require more value for each dollar they spend (Rust et al. 2015). Improving product value (better quality or quantity) and applying competitive pricing is the key to enhancing value equity (Rust et al. 2004). As mentioned, digital transformation facilitates product quality improvements, benefits, and features (Huang and Rust 2021). It further helps SMEs to improve customer experience and reduce costs necessary for competitive pricing (Teng and Yang 2022).

Regarding brand equity (BE), it suggests brand awareness, brand image, perceived quality, and brand loyalty (Aaker 1996). BE signalises the brand's strength and indicates how strong, clear, and unique the brand is in customers' minds, which can be enhanced through digital marketing practices (Niros and Niros 2023). As a result, effective digital marketing efforts result in the communication of strong brand identity, brand cues, and *one-to-one direct digital communication* consistent with all the touchpoints in authentic and meaningful ways (Kotler et al. 2017). Hence, SMEs with a successful digital transformation are capable of instigating customer loyalty, quality, and brand awareness and communicate a unique image as a means of differentiation and competitive advantage.

Last, but not least, relationship equity (RE) concerns the emotion-based connection customers develop with a brand that is triggered by strong and enduring relationships (Rust et al. 2004). The customer-brand relationship can be cultivated and enhanced via digital tools and technologies enabling SMEs to provide brand-related cues through virtual reality, interact with numerous customers on a personal level, and *establish positive and memorable customer experiences related to customer care service* (Niros and Niros 2023; Teng and Yang 2022). Consequently, SMEs can apply the RE strategy that focuses on improving customer retention and customer lifetime value through solid relationships and sentiments of loyalty and trust enduring in time. This type of connection further develops positive word of mouth and referrals that are particularly important for SMEs' further growth (Melo et al., 2023).

Summarising, the integration of digital transformation (DT) in marketing mix and strategy implementation is particularly important and beneficial, especially in today's turbulent environment. Digital technologies can be used to enhance organisational and marketing performance through improved customer experience, reach, decision-making based on real- time information, and one-to-one customer communication that results in brand preference at low cost. Hence, SMEs are more prone to adjust to the demands of the digital age and gain a competitive advantage in a VUCA business landscape, thus developing marketing agility. Figure 2 below represents the development of marketing agility as a result of DT and its respective benefits on marketing efforts and strategies that positively impact brand experience and CLV. As mentioned previously, the digital transformation process needs to consider the involvement of consumers to ensure social and economic sustainability for SMEs. Although DT into all aspects of marketing strategies and activities provides unparalleled opportunities for SMEs to enhance marketing efficiency, engage with customers, and increase the level of experience provided (Spita et al., 2020), it also poses significant perils in terms of consumer-related concerns making DT efforts not always reaching their goals. Thus, it is essential for SMEs to consider the possible perils that consumers may encounter to ensure a successful implementation of DT in marketing. Next, this section explores the customer-related perils associated with SMEs' DT in marketing.

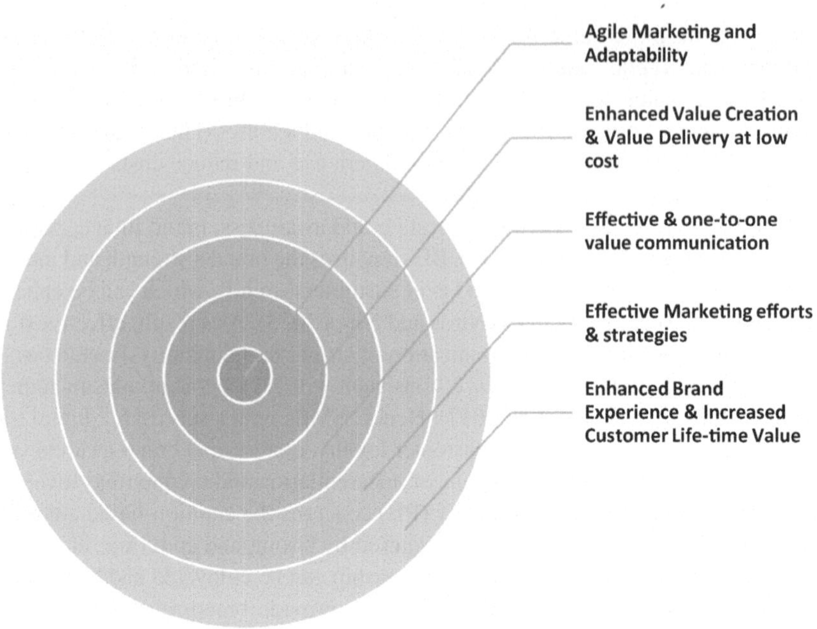

Fig. 2. Benefits of DT in Marketing: From Agile Marketing to increased CLV

4 The Perils of Digital Transformation in Marketing from a Consumer Perspective

While SMEs face numerous challenges when implementing DT, such as limited corporate capabilities and employee resistance, they also face several challenges and potential risks deriving from consumer skepticism, significantly affecting or even threatening the successful implementation of SMEs' DT efforts. Consumer skepticism, negative cognition towards DT technologies, and their reluctance or inability to accept and use them can be attributed to several but interlaced factors (see Fig. 3 below), including consumers' mistrust of the capabilities of SMEs to drive the DT, increasing consumer concerns over privacy and data security, consumer concerns about the social impact of the SMEs' transition to DT, concerns about the complexity of DT technologies, and the potential loss of personal contact in their transactions.

Peril 1: Consumers' General Technological Skepticism and Technological Fatigue
Consumers frequently express concerns about the negative consequences of technology, such as excessive reliance and limited or no control over it (Parasuraman 2000) e.g., customers are concerned about addiction to smartphones or social media usage and they do not always perceive AI-based services as a positive experience (Sanz-Blas et al. 2019; Kaplan and Haenlein 2020). As a result, negative feelings may be evoked and consumers may experience technological anxiety and even technophobia (Shu et al. 2011), feel discomfort and uneasiness when using AI-based services (Flavián et al. 2022), or even fear of using AI and robots (Mende et al. 2019). These psychological barriers

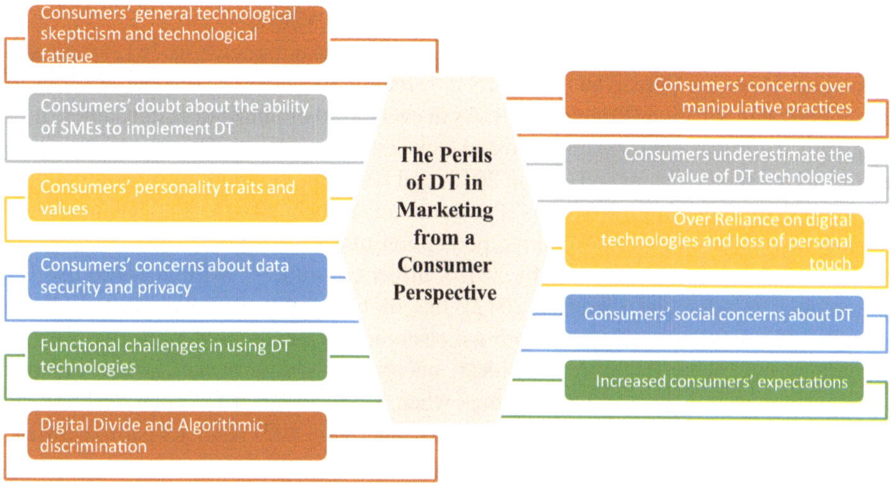

Fig. 3. The Perils of Digital Transformation (DT) in Marketing from a Consumer Perspective

are becoming increasingly pronounced and directly influence consumers' skepticism and reluctance to accept and adopt DT technologies, e.g., technological anxiety negatively affects chatbot usage (Kwangsawad and Jattamart 2022).

Furthermore, DT can also be negatively affected by consumers' digital overload and the subsequent negative effect on their physical health. Consumers often feel exhausted due to their prolonged exposure to DT initiatives, e.g., this fatigue is evident in social media (Zheng and Ling 2021, Bright et al. 2022) and digital advertising (Silberstein et al. 2023), leading to reduced engagement and increased resistance to these technologies. Moreover, personalized ads and digital communication also contribute to consumers' digital fatigue.

Peril 2: Consumers Doubt the Ability of SMEs to Implement DT

While some consumers distrust technology in general, others are skeptical and insecure about its ability to work properly. This may arise from their perception that SMEs experience some difficulties in their journey to DT, including organizational, financial, and technological constraints (Elhusseiny and Crispim 2022), due to their often low or inadequate digital capabilities, competencies and skills, particularly when compared to larger, well-established organizations. Increased consumers' negative perception about the ability of SMEs to deliver valuable results through DT technologies, will increase their perceived risk and negatively influence their willingness to use them (Roy et al. 2018). Negative consumer perceptions can be heightened by consumers' previous negative experiences with an SME's DT technology that improperly operated in the past, failing to deliver the expected value which adversely affected customers' experience and satisfaction. This may lead to perceptions of increased performance risks if SME's technology is used again in the future, e.g. in the use of self- service technology (Le et al. 2020).

Peril 3: Consumers' Personality Characteristics and Values

Given that the applications of DT technologies in marketing such as IoT, smart services, and robots are perceived as new by consumers, they can be considered forms of technological innovation (Rogers 1995). As in every kind of innovation, consumers may frequently exhibit a general skepticism and lack of trust towards these new technologies, leading to resistance (Mani and Chouk 2018) and late adoption. Initially, an individual's level of technological innovativeness (Parasuraman and Colby 2015), and technological optimism, which refers to the positive perception of technology being functional and trustworthy (Parasuraman and Colby 2015), affect consumers' willingness to use DT technologies (Godoe and Johansen 2012), like mobile payment (Oliveira et al. 2016) and robo-advisors. Innovative and optimist customers typically hold a positive view of technology, even when its future benefits are uncertain (Prodanova et al. 2018), and may also overlook possible negative outcomes when trying to adapt to a new technology. Customers who lack these characteristics (Reis and Melão 2021) may jeopardize SMEs' efforts towards DT.

Moreover, it is also expected that some consumers will exhibit higher levels of skepticism and resistance to new or unfamiliar DT innovations in marketing (Mani and Chouk 2018; Hajiheydari et al. 2021), as they create major changes in consumers' way of interacting with brands, may disrupt their routine or may conflict with existing beliefs and values (Ram and Sheth 1989). DT technologies can transform the way consumers function and behave, which can influence their attitudes toward new technologies (Yu and Chantatub 2016) and provoke negative emotional responses, leading to heightened ambiguity and risk over potential negative outcomes, as well as upsetting their sense of stability. Thus, consumers with traditional beliefs, preference for stability, and intolerance towards anything that challenges the status quo (Heidenreich et al. 2022) have additional psychological barriers (Antioco and Kleijnen 2010) and expected to be more resistant to accepting DT technologies that trigger change, e.g. resistance to using online chatbots (Jan et al. 2023) or Mobile Payment Systems which negate customers' existing routine by requiring them to conduct digital payments instead of conventional cash-based payments, while status quo bias positively affects resistance towards the use of AI voice assistants (Balakrishnan et al. 2024; Heidenreich and Spieth 2013).

Peril 4: Functional Challenges in Using DT Technologies

Although DT technologies have sophisticated capabilities and have the potential to offer a seamless and enhanced customer experience, customers often experience technical difficulties and perceive them as difficult to understand, inconvenient, or complex to use. Some technologies like IoT are already mature and widely accepted by consumers, while others more innovative ones, including blockchain and AI-based applications, encompass complex and sophisticated technical characteristics, requiring advanced technical skills or cognitive effort by consumers to use them. Functional barriers, such as the perceived level of complexity and the amount of cognitive effort required for learning and using a certain technology, are frequently encountered by particularly low-tech-skilled consumers. For example, consumers lacking technological proficiency may face challenges in adopting novel digital payment systems like bitcoins, due to their perceived complexity (Leung and Dickinger 2017).

In addition, technology has enabled companies to augment the capabilities of their products with additional features, as in the case of smart products. Paradoxically, an excess of features can decrease a product's usability, rendering it complex and difficult to use. Consumers may feel overwhelmed by high-featured products, for example when interacting with over- specified smart products (Cesaretto et al. 2021), leading to dissatisfaction, frustration, and cognitive strain with the product's unnecessary features (Thompson et al. 2005).

These functional and technical difficulties act as barriers to the acceptance of DT technologies and can explain customers' resistance to them (Laukkanen 2016), as they can't integrate with them efficiently and reap all the relevant benefits, leading to poor customer experience and satisfaction, which can make them unable to continue using these technologies (Mani and Chouk 2017).

Peril 5: Consumers' Concerns about Data Security and Privacy

Many technologies require consumers to disclose personal information to use their services through an online chatbot or smartphone app, and consumers usually contribute their data in exchange for using these services (Acquisti et al. 2020; Kokolakis 2017). By leveraging big data through analytics applications, SMEs have the opportunity to decipher consumer behavior and preference patterns and provide highly personalised recommendations, customise their marketing offerings and experiences to their preferences, and deliver targeted content and campaigns. However, customers often are hesitant about giving their consent to utilize their personal data due to data security and privacy issues.

The management of substantial amounts of sensitive customer data for personalisation purposes (Grover et al. 2018), such as customers' phone numbers and location data, have also spawned many new heightened risks. Data security issues pose one of the most significant ones that may hinder the benefits of DT in marketing, as customers exhibit a heightened awareness of risks associated with using DT technologies regarding information security and their trust and loyalty largely depend on how the company protects their data. For example, consumers usually raise concerns about the exposure of their financial data in Mobile Payment Systems (Laukkanen 2016), while insecurity concerns reduce consumers' willingness to use robo-advisors (Roh et al. 2023). Cybersecurity concerns also present significant threats for SMEs.

Apart from jeopardising customer data, customers are becoming increasingly aware of their data privacy rights to know that their data is handled by different companies within the big data value chain without having ownership rights over them (Rajaobelina et al. 2021). Consumers who share their data often grapple with concerns regarding how their personal information is being collected, used, and protected and may feel that their personal data is misused, being leaked from Big Data databases, or sold to other companies, without their consent. In this context, inflated concerns around data privacy and data misuse also represent an ethical challenge for SMEs engaging in DT. For example, customers using chatbots worry about their information being recorded and not securely used (Pillai et al. 2024; Chang and Hsiao 2024). There are also significant concerns over consumer privacy in push notification service records (Cheng et al. 2022) and data tracking to optimise omnichannel marketing (Cui et al. 2019). Customers' perceived psychological risk (Marafon et al. 2018) and the sense of distrust between customers and the company or brand enhance customers' resistance to using DT technologies, e.g., concerns regarding information privacy have been found to influence consumers' resistance to smart services (Mani and Chouk 2019).

At last, all the above customers' rising privacy and insecurity concerns coupled with their skepticism to divulge personal information, not only minimize the benefits SMEs can gain from the utilisation of data to predict customer preferences (Carmon et al. 2019) but also elevate SMEs' costs as they address data protection strategies and potential legal issues.

Peril 6: Consumers' Concerns Regarding Digital Divide and Algorithmic Discrimination

DT can exacerbate existing inequalities in consumers' access to DT technologies and create new forms of discrimination. Since consumers possess varying levels of technological literacy, many have limited technological proficiency and experience or may lack access to digital tools, DT can inadvertently deepen the digital divide by reinforcing the privilege of some consumers having access to DT technologies while excluding others from taking full advantage of DT-related benefits, particularly older ones and consumers in less developed countries. In addition, the digital divide can be due to consumers' different financial statuses, presenting a significant challenge for SMEs. For example, wealthier consumers can afford to opt for premium ad-free apps and services (Zennyo 2020), while those with lower incomes who cannot afford additional subscription fees and continue to be exposed to personalised but also annoying targeted ads, leading to limited control over their privacy preferences (Taylor 2024).

An additional peril that SMEs may encounter in their DT is further consumer discrimination due to socially biased algorithms utilised in their digital marketing models (Benlian et al. 2018). Algorithms allow SMEs to personalise advertisements and develop targeted multi- channel campaigns (Vermeer et al. 2019), while making personalised product recommendations based on consumers' behaviour and preferences (Davenport et al. 2020; Kumar et al. 2020). Machine learning algorithmic biases occur, intentionally or unintentionally, when algorithm decisions or predictions reflect prejudice or favouritism towards specific customers or customer groups related to gender, age, race, sexual orientation, or other factors (Lambrecht and Tucker 2019; O'Neil 2016), leading to biased, unjust, or discriminatory outcomes (Kiener 2020). Inadequate algorithmic designs, limited representation of specific consumer groups in the datasets used to train AI algorithms, low quality of training data (Favaretto et al. 2019; Israeli and Ascazra 2022, Akter et al. 2021) or the lack of diversity among AI developers (Nouri 2021) usually contribute to biased judgments in AI-based marketing models.

Examples include Apple card's discriminatory credit limits policy against women, Amazon's Prime discriminatory pricing based on customers' location (Ingold and Soper 2016), Orbitz's pricier hotel suggestions to Mac users (Mattila and Choi 2014), and Uber's expensive pricing for certain destinations based on African-American population density (Pandey and Caliskan 2021). As a result, many consumers are either excluded or receive unfair outcomes, e.g. inferior levels of service, pricing, and promotional activities. Systematically disadvantaging or prohibiting certain groups of customers from equitable access to marketing offerings may result in adverse consequences for brands, such as reputational risks and consumer distrust (Srinivasan and Sarial-Abi 2021; Ukanwa et al. 2021).

Peril 7: Consumers' Concerns about Manipulative Practices

The extensive use of DT information technologies enables SMEs to collect, and analyse data to gain a deeper understanding of consumers' attitudes and behaviour that enables the identification of different market segment motivations and weaknesses, e.g., the analysis of transaction data from bank card payments (Budanov and Aseeva 2019). However, the increased insight into consumer behaviour has raised consumer concerns about whether this information can be used to influence their decision-making behaviour (Susser et al. 2019) through persuasion, deception, or nudging (de Marcellis-Warin et al. 2022). Furthermore, regarding the increasing use of algorithms, such as pricing and recommendation algorithms, to create personalized offerings, consumers, rather than being exposed to all available options and choices in the market, may be excluded, challenging their autonomy and freedom of choice (Zuboff 2015).

Peril 8: Consumers Underestimate the Value of DT Technologies

Although SMEs leverage DT technologies into their marketing strategies, such as augmented reality cabins in hotels, chatbots for better service, or guided visits in museums, customers may not always understand, appreciate, and often underestimate the value and potential benefits of using these technologies, e.g. travel consumers lack awareness of blockchain benefits (Önder and Treiblmaier 2018). Some new technologies require specific knowledge or experience that less tech-savvy consumers may lack. Additionally, the time and effort required to learn such technologies may not immediately translate into economic or non- economic benefits, resulting in their acceptance to backfire. Consumers with low-value awareness of a certain technology, due to a lack of relevant information or experience, are expected to resist using this technology (Chang and Hsiao 2024; Mani and Chouk 2018; Talwar et al. 2020), as the perception of lower value may elicit negative experiences and emotions (Alamäki et al. 2021).

Peril 9: Over-Reliance on Technology and Lack of Personal Touch

Many SMEs, especially in service industries, have adopted automation technologies to enable tasks to be carried out without human involvement, for example using hotel service robots for reception services, online customer service assistance chatbots, and AI agents that monitor and assist salespersons during tele-conversations. Despite consumers' increasing familiarity with technology in their everyday lives, human interaction remains important for customers in service encounters (Evanschitzky et al. 2015). Even though these technologies have sophisticated capabilities, such as voice analysis to detect and recognise consumers' emotions based on their tone, they lack empathy and understanding compared to human employees and also generate impersonal and generic reactions.

Customers who prefer interacting with humans may perceive anthropomorphic AI devices as incapable of human-like care and, as a result, they may be less inclined to use them (Kaur and Gupta 2012) or perceive a service failure, e.g., customers do not prefer AI-based service chatbots over employees during service encounters (Liu-Thompkins et al. 2022). This perceived lack of human touch and empathy in AI-based service chatbots can trigger unintended negative emotions from customers, such as disappointment and dissatisfaction.

(Bo Yang et al 2023). Customers may also feel uneasy with AI monitoring conversations (Davenport et al. 2020), and discomfort with humanoid robots (Mende et al. 2019), while they can become frustrated with mechanical algorithms (Yam et al. 2021; Wieseke et al. 2012). Thus, incorporating AI in service settings may not always enhance the brand experience, since the impersonal nature of digital interactions might negatively impact customers' affective experience (Kelly et al. 2023; Chan and Tung 2019) and make them feel detached or less emotionally connected to the organization/brand.

Peril 10: Consumers' Social Concerns about DT
DT in marketing and the integration of emerging technologies are reshaping the practice of marketing and profoundly changing the workplace. DT technologies have automated many routine tasks, while the deployment of advanced technologies, such as virtual assistants, chatbots, robotics, AR and VR interfaces, or branded avatar interactions, have made them also capable of handling analytical tasks. This has raised consumers' concerns about the possible adverse impact of DT on the labor market, potentially leading to job losses and job displacement (Dengler and Gundert 2021), especially for occupations that are more susceptible to automation (Plumwongrot and Pholphirul 2022). Additionally, they may also raise concerns about the environmental impact of DT, such as the excess energy consumption of data centers.

Peril 11: Increased Consumers' Expectations
Currently, many SMEs dealing with a diverse consumer base are constantly utilising novel digital technologies, such as AI, to deliver customised consumer-focused experiences (Huang and Rust 2021). Consequently, customers' expectations for technological capabilities are increasing (Sestino et al. 2020) largely driven by these technological advancements. Consumers expect companies engaging in DT initiatives to incorporate sophisticated technologies for personalised experiences that fully address their expectations. For example, customers interacting with brands across various digital and offline channels and devices have high expectations, which makes effectively managing omnichannel offerings across different touchpoints and also aligning them with customer expectations increasingly complex (Cui et al. 2021; Natarajan and Veera Raghavan 2023).

In this context, SMEs are facing increasing pressure in their DT efforts, as they often struggle to keep up with the inflated expectations of their customers. SMEs that underestimate this peril and fail to align DT strategies with customer expectations and needs are at risk of facing weaker customer relationships, lower customer engagement, and decreased satisfaction (Diener and Špaček 2021). This could eventually lead to losing customers to more agile competitors (Matzler et al. 2018). For example, the widespread use of AI-based chatbots often fails to meet customer expectations for human-like interactions (Adam et al. 2021), as chatbots typically provide generic and unsuitable responses to customer inquiries (Aslam 2023), leading to unsatisfactory encounters with chatbots.

5 Conclusion

In the chapter's *first research question*, we explored the digital technologies SMEs can use to initiate the digital transformation process in marketing. Digital technologies such as artificial intelligence, machine learning, big data analytics, and other technological advancements (i.e. virtual technology) suggest some of the most sophisticated enablers of digital transformation in marketing, especially for SMEs. As for *the second research question,* we unveiled the benefits of these digital technologies. DT facilitates SMEs in automating tasks, customising customer experience throughout the various touchpoints, proceeding with data-driven decision-making, and pushing them toward revenue and profit growth via effective marketing strategies and practices.

Regarding the *third research question*, we illustrated the perils of Digital Transformation in Marketing from a Consumer Perspective for SMEs. Despite benefits, marketers need to acknowledge the risks associated with DT in marketing, including data privacy and security concerns, lack of personal interaction, digital divide, exclusion of consumers, and increased customer expectations, as well as the severity and impact of these perils. The effect of the aforementioned perils can significantly impact customer experience, consumer trust, brand loyalty, and brand image and minimise the benefits of DT. Lastly, concerning the *fourth research question*, we propose specific actions for managers and policy makers to minimise the impact of consumer-related perils in SMEs.

5.1 Theoretical Implications

Even though many pieces of research stress the importance of marketing technologies and digital transformation in marketing performance, to the best of our knowledge, few papers have explored the benefits in tactical and strategic terms and how they result in brand, relationship and value equity in SMEs. The theoretical implications for Research questions 1 and 2 suggest that the integration of marketing technologies in the process of DT is vital for SMEs to develop agile marketing practices and strategies, bolster customer lifetime value (CLV), and simultaneously achieve higher levels of brand experience that strengthen the link between marketing practices, customer perceptions and brand preference.

Although many research findings posit that DT is challenging for several different issues and concerns, considerably less work has been undertaken on deploying the factors that may explain the overall customer skepticism towards DT in SMEs and addressing research questions 3 and 4. The review of DT literature shows that research in this field mainly focuses on specific consumer-related perils of DT at a time, rather than considering overall consumer skepticism. We add to this line of research by presenting a checklist of challenges that might impede DT success from a consumer perspective, offering a novel contribution to our understanding of DT acceptance.

Given the divide in consumers' attitudes, it would seem reasonable that many of the customers' concerns towards DT will differ. A central contribution of this study is to provide a framework for future researchers to examine more thoroughly the variance in consumer attitudes towards various aspects of DT in a holistic manner, and far from merely focusing on classic aspects of technological skepticism. This framework can guide future research to evaluate which aspects of consumer skepticism are more prominent in certain situations and which actions SMEs should take to gain customer trust in this process.

5.2 Managerial Implications

This chapter suggests a practical guide for managers and entrepreneurs for SMEs to achieve faster growth and adaptability to today's turbulent business environment, characterised by volatility, uncertainty, complexity and ambiguity through digital transformation. The application of DT in the marketing function results in agility and higher efficiency that answer the first two research questions. First and foremost, entrepreneurs should instil and develop a market orientation culture before proceeding on DT. On the other hand, they need to invest highly in human resources and new processes and, of course, in new digital marketing technologies 4.0, such as artificial technology, big data analytics, and virtual technologies. As a result of DT, they enhance the brand experience, brand-related cues, perceived quality, brand loyalty, brand image, consumer-brand relations, and perceptions of the value-for-money necessary to withstand fierce competition.

Concerning the last two research questions, managers and policy makers need to take some actions on the perils unveiled. DT is challenging and SMEs should understand and carefully identify why consumers resist or mistrust DT technologies, evaluate potential risks associated with DT, and find ways to overcome barriers which eventually will facilitate the adoption process of DT technologies from consumers. SMEs must invest in encouraging skeptical consumers to overcome their resistance toward the acceptance of these technologies. Regarding consumers' general technological skepticism, SMEs should consider and address these issues by engaging in transparent communication with customers regarding the benefits of using their DT technologies. In addition, addressing digital fatigue when trying to reach and communicate to fatigued audiences, SMEs should carefully estimate potential benefits and potential shortfalls of excessive repeated communication and personalized messages.

In the context of increased consumers' doubt of SMEs' ability to effectively run DT, SMEs should ensure that DT technologies offer reliable services, hence encouraging their acceptance. They must also consider technological difficulties and develop ways to overcome consumers' resistance behavior due to functional barriers. Also, SMEs targeting more conservative customers should be aware that those customers who value stability in their lives are challenging to reach through technologies that are not in congruence with their beliefs and values. Thus, SMES need to consider psychological variables when setting up DT technologies.

Additionally, marketers on their route to DT need to recognize the increased ethical implications of using large amounts of consumer data and increasingly sophisticated data analytics in their DT process and to address issues regarding the increasingly high levels of customers' privacy concerns effectively (Anton et al. 2010; Acquisti et al. 2020). SMEs should implement ethical and responsible practices when using and storing customer information (Strohbach et al. 2016), always complying with data protection regulations, while being transparent about how their data are collected and used (Cui et al. 2021; Cheah et al. 2022). In data-driven strategies, when it comes to consent giving, obtaining informed consent and building trust for consent is crucial (Hashem et al. 2015;

Ozbal et al. 2020), as trust can overcome consumers' privacy concerns (Evjemo et al. 2018).

SMEs undergoing DT should also address any potential digital gap by developing digital strategies that are inclusive of all consumers (Lythreatis et al. 2022) and ensure that DT initiatives do not deprive any customer group of access to different DT technologies, e.g., by providing alternative channels for consumer engagement. This approach can help SMEs expand their customer base and encourage DT technologies' use and acceptance. Additionally, given that customers are also increasingly concerned about whether AI algorithms incorporate human biases into their systems (Tsamados et al. 2021), SMEs should increase their comprehension of potential biases and cultivate expertise in preventing deep-rooted social biases within their machine-learning models (Puntoni et al. 2021) to ensure that they are employed equitably and inclusively.

Another implication for SMEs undergoing DT of marketing aiming to increase a DT technology's acceptance and utilization is to identify consumers' value barriers and effectively communicate valuable information about the added benefits of a new technological service, encourage awareness and perception of its usefulness, e.g., customers are more likely to use robo-advisors or mobile-payment solutions like Apple Pay when they become more conscious of a new technology's relative benefits and value (Flavian et al. 2022; Kaur et al. 2020). However, over-reliance on technology and digital tools and overlooking the significance of human interaction for customers is another major peril for SMEs undergoing DT. SMEs need to recognize that the human factor still remains important when digitally transforming (Magistretti et al. 2021). It's important for SMEs to carefully integrate human interaction in DT technologies and balance automation with human touchpoints to ensure optimal experiences for all customers (Bolton et al. 2018). Furthermore, in certain industries, such as luxury products and family businesses, personalized service is a key differentiator, and customers highly value the personal dynamics of human relationships in their service experience (Laukkanen and Kiviniemi 2010).

From a practical point of view, SMEs engaging in DT initiatives in marketing should also consider consumers' social concerns and address them accordingly by emphasizing that the use of DT technologies, including for marketing purposes, can also lead to improvements in labor market efficiency, as existing jobs may change rather than disappear, and new job opportunities may arise that require a new set of skills. Additionally, SMEs should also work toward communicating environmental sustainability initiatives.

Also, SMEs should consider their customers' evolving expectations when investing in DT for marketing. This will enable SMEs to create a customer-focused DT strategy, ensuring a balance between technology and consumer expectations, and ensuring that digital strategies resonate with them (Brill et al. 2019). Adapting to changing and increasing customer expectations also requires SMEs to be agile when implementing DT processes. Furthermore, SMEs should effectively manage customer expectations by clearly communicating what customers should expect before using a new technological system.

Lastly, SMEs can significantly benefit from segmenting their consumers based on the specific obstacles they encounter in the context of Digital Transformation. By identifying distinct segments, such as technologically skeptical customers and those who underestimate the value of DT, SMEs can tailor their strategies and communications to resonate with the unique challenges of each group.

To conclude, transparently communicating with customers about the benefits of DT throughout the transformation process, adopting a customer-centric approach and prioritizing ethical considerations can mitigate the risks of DT. Public policy can also significantly enhance SMEs' ability to navigate the complexities of digital transformation by addressing budgetary constraints and other training, infrastructural, and legal obstacles. Supporting SMEs in undertaking the above actions can benefit overall economic growth as well.

5.3 Limitations and Future Research

It would be interesting to study whether differences exist in consumers' concerns toward DT in different product categories. Also, a potentially exciting topic for future studies would be to examine the impact of each aspect of consumer scepticism toward SMEs' digital transformation on several outcomes related to consumers' attitudes and behaviours, such as purchase intentions and loyalty, among others, in different settings. Lastly, it would be essential to research the perils deriving from SMEs' employees when firms adopt new technologies in the process of DT.

References

Aaker, D.A.: Measuring brand equity across products and markets. Calif. Manag. Rev. **38**(3), 1–20 (1996)

Acquisti, A., Brandimarte, L., Loewenstein, G.: Secrets and likes: the drive for privacy and the difficulty of achieving it in the digital age. J. Consum. Psychol. **30**, 30736–30758 (2020). https://doi.org/10.1002/jcpy.1191

Adam, M., Wessel, M., Benlian, A.: AI-based chatbots in customer service and their effects on user compliance. Electron Markets **31**, 427–445 (2021). https://doi.org/10.1007/s12525-020-00414-7

Akter, S., et al.: Algorithmic bias in data-driven innovation in the age of AI. Int. J. Inf. Manag. **60**, 102387 (2021)

Alamäki, A., Dirin, A., Suomala, J.: Students' expectations and social media sharing in adopting augmented reality. Int. J. Inf. Learn. Technol. **38**(2), 196–208 (2021). https://doi.org/10.1108/IJILT-05-2020-0072

Antioco, M., Kleijnen, M.H.P.: Consumer adoption of technological innovations: effects of psychological barriers in a lack of content versus a presence of content situation. Eur. J. Mark. **44**(11/12), 1700–1724 (2010). https://doi.org/10.1108/03090561011079846

Anton, A.L., Earp, J.B., Young, J.D.: How Internet users' privacy concerns have evolved since 2002. Computer Science Technical Report TR-2009–16, North Carolina State University (2010)

Aslam, U.: Understanding the usability of retail fashion brand chatbots: Evidence from customer expectations and experiences. J. Retail. Consum. Serv. **74**, 103377 (2023)

Balakrishnan, J., Dwivedi, Y.K., Hughes, L., Boy, F.: Enablers and inhibitors of AI-powered voice assistants: a dual-factor approach by integrating the status quo bias and technology acceptance model. Inf. Syst. Front. **26**(3), 921–942 (2024)

Benlian, A., Kettinger, W.J., Sunyaev, A., Winkler, T.J.: Special section: the transformative value of cloud computing: a decoupling, platformization, and recombination theoretical framework. J. Manag. Inf. Syst. **35**(3), 719–739 (2018). https://doi.org/10.1080/07421222.2018.1481634

Yang, B., Sun, Y., Shen, X.L.: Understanding AI-based customer service resistance: a perspective of defective ai features and tri-dimensional distrusting beliefs. Inf. Process. Manag. **60**(3), 103257 (2023)

Bolton, R.N., et al.: Customer experience challenges: bringing together digital, physical and social realms. J. Serv. Manag. **29**(5), 776–808 (2018). https://doi.org/10.1108/JOSM-04-2018-0113

Bright, L.F., Logan, K., Lim, H.S.: Social media fatigue and privacy: an exploration of antecedents to consumers' concerns regarding the security of their personal information on social media platforms. J. Interact. Advert. **22**(2), 125–140 (2022). https://doi.org/10.1080/15252019.2022.2051097

Brill, T.M., Munoz, L., Miller, R.J.: Siri, Alexa, and other digital assistants: a study of customer satisfaction with artificial intelligence applications. J. Mark. Manag. **35**(15–16), 1401–1436 (2019). https://doi.org/10.1080/0267257X.2019.1687571

Budanov, V., Aseeva, I.: Manipulative marketing technologies in new digital reality. Economic Annals-XXI/Ekonomìčnij Časopis-XXI **180**(11/12), 58–68 (2019)

Carmon, Z., Schrift, R., Wertenbroch, K., Yang, H.: Designing AI systems that customers won't hate. MIT Sloan Manag. Rev. **61**(2), 1–6 (2019)

Cesaretto, R., Buratto, A., Giovanni, P.D.: Mitigating the feature fatigue effect for smart products through digital servitization. Comput. Ind. Eng. **156**, 107218 (2021)

Chan, A.P.H., Tung, V.W.S.: Examining the effects of robotic service on brand experience: the moderating role of hotel segment. J. Travel Tour. Mark. **36**(4), 458–468 (2019). https://doi.org/10.1080/10548408.2019.1568953

Chang, T.S., Hsiao, W.H.: Understand resist use online customer service chatbot: an integrated innovation resist theory and negative emotion perspective. Aslib J. Inf. Manag. (2024). https://doi.org/10.1108/AJIM-12-2023-0551

Cheah, J.-H., Lim, X.-J., Ting, H., Liu, Y., Quach, S.: Are privacy concerns still relevant? Revisiting consumer behaviour in omnichannel retailing. J. Retail. Consum. Serv. **65**, 102242 (2022)

Cioppi, M., Curina, I., Francioni, B., Savelli, E.: Digital transformation and marketing: a systematic and thematic literature review. Italian J. Mark. **2023**(2), 207–288 (2023)

Cui, H., et al.: Omnichannel marketing: the challenge of data-integrity. Northeastern U. D'Amore-McKim School of Business Research Paper No. 3460580, NYU Stern School of Business (2019). SRN: https://ssrn.com/abstract=3460580. https://doi.org/10.2139/ssrn.3460580

Cui, H., Xiao, L., Zhang, X.: Application of mobile big data and artificial intelligence in the efficiency of e-commerce industry. Mob. Inf. Syst. **2021**, 4825643 (2021). https://doi.org/10.1155/2021/4825643

Davenport, T., Guha, A., Grewal, D., Bressgott, T.: How artificial intelligence will change the future of marketing. J. Acad. Mark. Sci. **48**(1), 24–42 (2020)

De Bruyn, A., Viswanathan, V., Beh, Y.S., Brock, J.K.U., Von Wangenheim, F.: Artificial intelligence and marketing: pitfalls and opportunities. J. Interact. Mark. **51**(1), 91–105 (2020)

de Marcellis-Warin, N., Marty, F., Thelisson, E., et al." Artificial intelligence and consumer manipulations: from consumer's counter algorithms to firm's self-regulation tools. AI Eth. **2**, 259–268 (2022). https://doi.org/10.1007/s43681-022-00149-5

Dengler, K., Gundert, S.: Digital transformation and subjective job insecurity in Germany. Eur. Sociol. Rev. **37**(5), 799–817 (2021). https://doi.org/10.1093/esr/jcaa066

Diener, F., Špaček, M.: Digital transformation in banking: a managerial perspective on barriers to change. Sustainability (2021)

Elhusseiny, H.M., Crispim, J.: SMEs, barriers and opportunities on adopting industry 4.0: a review. Procedia Comput. Sci. **196**, 864–871 (2022)

Evanschitzky, H., Iyer, G.R., Pillai, K.G., Kenning, P., Schütte, R.: Retail service innovation. J. Prod. Innov. Manag. **32**, 459–475 (2015). https://doi.org/10.1111/jpim.12241
Evjemo, B., Castejón-Martínez, H., Akselsen, S.: Trust trumps concern: findings from a seven-country study on consumer consent to 'digital native' vs. 'digital immigrant' service providers. Behav. Inf. Technol. **38**(5), 503–518 (2018). https://doi.org/10.1080/0144929X.2018.1541254
Favaretto, M., De Clercq, E., Elger, B.S.: Big data and discrimination: perils, promises and solutions: a systematic review. J. Big Data **6**(1), 12 (2019). https://doi.org/10.1186/s40537-019-0177-4
Flavián, C., Pérez-Rueda, A., Belanche, D., Casaló, L.V.: Intention to use analytical artificial intelligence (AI) in services – the effect of technology readiness and awareness. J. Serv. Manag. **33**(2), 293–320 (2022). https://doi.org/10.1108/JOSM-10-2020-0378
Gillpatrick, T.: The digital transformation of marketing: Impact on marketing practice & markets. Econ.-Innov. Econ. Res. J. **7**(2), 139–156 (2019)
Gobble, M.M.: Digital strategy and digital transformation. Res. Technol. Manag. **61**(5), 66–71 (2018)
Godoe, P., Johansen, T.S.: Understanding adoption of new technologies: technology readiness and technology acceptance as an integrated concept (2012)
Grover, V., Chiang, R.H.L., Liang, T.P., Zhang, D.: Creating strategic business value from big data analytics: a research framework. J. Manag. Inf. Syst. **35**(2), 388–423 (2018). https://doi.org/10.1080/07421222.2018.1451951
Hajiheydari, N., Delgosha, M.S., Olya, H.: Scepticism and resistance to IoMT in healthcare: application of behavioural reasoning theory with configurational perspective. Technol. Forecast. Social Change **169**, 120807 (2021). ISSN 0040–1625
Hashem, I.A.T., Yaqoob, I., Anuar, N.B., Mokhtar, S., Gani, A., Khan, S.U.: The rise of "big data" on cloud computing: review and open research issues. Inf. Syst. **47**, 98–115 (2015)
Heidenreich, S., Spieth, P.: Why innovations fail — the case of passive and active innovation resistance. Int. J. Innov. Manag. (IJIM) **17**(5), 1–42 (2013)
Heidenreich, S., Freisinger, E., Landau, C.: The dark side of business model innovation: an empirical investigation into the evolvement of customer resistance and the effectiveness of potential countermeasures. J. Prod. Innov. Manag. **39**(6), 824–846 (2022)
Huang, M.H., Rust, R.T.: A strategic framework for artificial intelligence in marketing. J. Acad. Mark. Sci. **49**, 30–50 (2021)
Ingold, D., Soper, S.: Amazon Doesn't Consider the Race of Its Customers. Should It?. Bloomerg (2016). https://www.bloomberg.com/graphics/2016-amazon-same-day/
Israeli, A., Ascarza, E.: Algorithmic Bias in Marketing. Harvard Business School Technical Note 521–020 (2022). Accessed July 2022
Jan, I.U., Ji, S., Kim, C.: What (de) motivates customers to use AI-powered conversational agents for shopping? The extended behavioral reasoning perspective. J. Retail. Consum. Serv. **75**, 103440 (2023)
Kaplan, A.M., Haenlein, M.: Rulers of the world, unite! the challenges and opportunities of artificial intelligence. Bus. Horiz. **63**, 37–50 (2020)
Kaur, G., Gupta, S.: Business orientation of indian consumer banking. Glob. Bus. Rev. **13**, 481–507 (2012)
Kaur, P., Dhir, A., Singh, N., Sahu, G., Almotairi, M.: An innovation resistance theory perspective on mobile payment solutions. J. Retail. Consum. Serv. **55**, 102059 (2020)
Kelly, S., Kaye, S.-A., Oviedo-Trespalacios, O.: What factors contribute to the acceptance of artificial intelligence? A systematic review. Telemat. Inf. **77**, 101925 (2023)
Kiener, M.: Artificial intelligence in medicine and the disclosure of risks. AI Soc. **36**, 1–9 (2020)
Kokolakis, S.: Privacy attitudes and privacy behaviour: a review of current research on the privacy paradox phenomenon. Comput. Secur. **64**, 122–134 (2017)

Kotler, P., Kertajaya, H., Setiawan, I.: Marketing 4.0 Moving From Traditional to Digital. John Wiley & Sons, Inc., Hoboken (2017)
Kumar, A., Shankar, R., Aljohani, N.R.: A big data driven framework for demand- driven forecasting with effects of marketing-mix variables. Ind. Mark. Manag. **90**, 493–507 (2020)
Kwangsawad, A., Jattamart, A.: Overcoming customer innovation resistance to the sustainable adoption of chatbot services: a community-enterprise perspective in Thailand. J. Innov. Knowl. **7**, 102211 (2022)
Lambrecht, A., Tucker, C.: Algorithmic bias? an empirical study of apparent gender-based discrimination in the display of STEM career ads. Manag. Sci. **65**(7), 2966–2981 (2019)
Laukkanen, T.: Consumer adoption versus rejection decisions in seemingly similar service innovations: the case of the internet and mobile banking. J. Bus. Res. **69**(7), 2432–2439 (2016)
Laukkanen, T., Kiviniemi, V.: The role of information in mobile banking resistance. Int. J. Bank Mark. **28**(5), 372–388 (2010). https://doi.org/10.1108/02652321011064890
Le, N.T., Rao Hill, S., Troshani, I.: Perceived control and perceived risk in self- service technology recovery. J. Comput. Inf. Syst. **62**(1), 164–173 (2020). https://doi.org/10.1080/08874417.2020.1756533
Leung, D., Dickinger, A.: Use of Bitcoin in Online Travel Product Shopping: The European Perspective. Information and Communication Technologies in Tourism (2017)
Liu-Thompkins, Y., Okazaki, S., Li, H.: Artificial empathy in marketing interactions: bridging the human-AI gap in affective and social customer experience. J. Acad. Mark. Sci. **50**, 1198–1218 (2022). https://doi.org/10.1007/s11747-022-00892-5
Lythreatis, S., Singh, S.K., El-Kassar, A.N.: The digital divide: a review and future research agenda. Technol. Forecast. Social Change **175**, 121359 (2022). https://doi.org/10.1016/j.techfore.2021.121359
Magistretti, S., Anh, C.T., Pham, C.D.: Enlightening the dynamic capabilities of design thinking in fostering digital transformation. Ind. Mark. Manag. **97**, 59–70 (2021). https://doi.org/10.1016/j.indmarman.2021.06.014
Mani, Z., Chouk, I.: Drivers of consumers' resistance to smart products. J. Mark. Manag. **33**(1–2), 76–97 (2017)
Mani, Z., Chouk, I.: Consumer resistance to innovation in services: challenges and barriers in the internet of things era. J. Prod. Innov. Manag. **35**(5), 780–807 (2018). https://doi.org/10.1111/jpim.12463.hal-03700875
Mani, Z., Chouk, I.: Impact of privacy concerns on resistance to smart services: does the 'Big Brother effect' matter? J. Mark. Manag. **35**(15–16), 1460–1479 (2019). https://doi.org/10.1080/0267257X.2019.1667856
Marafon, D.L., Basso, K., Espartel, L.B., de Barcellos, M.D., Rech, E.: Perceived risk and intention to use internet banking: the effects of self-confidence and risk acceptance. Int. J. Bank Mark. **36**(2), 277–289 (2018). https://doi.org/10.1108/IJBM-11-2016-0166
Matt, D.T., Rauch, E.: SME 4.0: the role of small-and medium-sized enterprises in the digital transformation. In: Industry 4.0 for SMEs: Challenges, Opportunities and Requirements, pp. 3–36 (2020)
Mattila, A.S., Choi, C.: An analysis of consumers' reactions to travel websites' discrimination by computer platform. Cornell Hosp. Q. **55**(2), 210–215 (2014). https://doi.org/10.1177/1938965513509879
Matzler, K., Friedrich von den Eichen, S., Anschober, M., Kohler, T.: The crusade of digital disruption. J. Bus. Strateg. **39**(6), 13–20 (2018). https://doi.org/10.1108/JBS-12-2017-0187
Melo, I.C., Queiroz, G.A., Junior, P.N.A., de Sousa, T.B., Yushimito, W.F., Pereira, J.: Sustainable digital transformation in small and medium enterprises (SMEs): a review on performance. Heliyon **9**(3), e13908 (2023)

Mende, M., Scott, M.L., van Doorn, J., Grewal, D., Shanks, I.: Service robots rising: how humanoid robots influence service experiences and elicit compensatory consumer responses. J. Mark. Res. **56**(4), 535–556 (2019). https://doi.org/10.1177/0022243718822827

Melanthiou, Y., Vasic, S.:. The metaverse in an era of volatility, uncertainty, complexity, and ambiguity (VUCA). In: Zarkada, A. (eds) Marketing Solutions to the Challenges of a VUCA Environment. GMA-GAMMA 2023. Springer Proceedings in Business and Economics. Springer, Cham (2024). https://doi.org/10.1007/978-3-031-58429-9_9

Miklosik, A., Evans, N.: Impact of big data and machine learning on digital transformation in marketing: a literature review. IEEE Access **8**, 101284–101292 (2020)

Niros, M.I., Niros, A.: Marketing solutions to the challenges of a volatile, uncertain, complex and ambiguous (VUCA) environment: applications in international marketing. In: GMA-GAMMA Joint Symposium, pp. 193–215. Springer, Cham (2023)

O'Neil, C.: Weapons of Math Destruction: How Big Data Increases Inequality and Threatens Democracy. Crown Publishing Group (2016)

Oliveira, H.C., et al.: Feeding behavior of lambs fed castor meal. Semin. Cienc. Agrar. **37**(3), 1451–1460 (2016)

Önder, I., Treiblmaier, H.: Blockchain and tourism: three research propositions. Ann. Tour. Res. **72**, 180–182 (2018). https://doi.org/10.1016/j.annals.2018.03.005

Ozbal, O., Duman, T., Topaloglu, O.: A trust-based peer-to-peer digital brand equity (P2P-DBE) model. J. Mark. Theory Pract. **28**(4), 497–520 (2020). https://doi.org/10.1080/10696679.2020.1794901

Pandey, A., Caliskan, A.: Disparate impact of artificial intelligence bias in ridehailing economy's price discrimination algorithms. In: Proceedings of the 2021 AAAI/ACM Conference on AI, Ethics, and Society (AIES '21), pp. 822–833. Association for Computing Machinery, New York (2021). https://doi.org/10.1145/3461702.3462561

Parasuraman, A.: Technology readiness index (TRI) a multiple-item scale to measure readiness to embrace new technologies. J. Serv. Res. **2**, 307–320 (2000). https://doi.org/10.1177/109467050024001

Parasuraman, A., Colby, C.L.: An updated and streamlined technology readiness index: TRI 2.0. J. Serv. Res. **18**(1), 59–74 (2015). https://doi.org/10.1177/1094670514539730

Pillai, R., Ghanghorkar, Y., Sivathanu, B., Algharabat, R., Rana, N.P.: Adoption of artificial intelligence (AI) based employee experience (EEX) chatbots. Inf. Technol. People **37**(1), 449–478 (2024). https://doi.org/10.1108/ITP-04-2022-0287

Plumwongrot, P., Pholphirul, P.: Are Robots stealing jobs? Empirical evidence from 10 developing countries. Econ. Innov. New Technol. **32**(6), 873–889 (2022). https://doi.org/10.1080/10438599.2022.2051020

Prodanova, J., San-Martín, S., Jimenez, N.: Are you technologically prepared for mobile shopping? Serv. Ind. J. **41**(9–10), 648–670 (2018). https://doi.org/10.1080/02642069.2018.1492561

Puntoni, S., Reczek, R.W., Giesler, M., Botti, S.: Consumers and artificial intelligence: an experiential perspective. J. Mark. **85**(1), 131–151 (2021). https://doi.org/10.1177/0022242920953847

Rajaobelina, L., Prom Tep, S., Arcand, M., Ricard, L.: The relationship of brand attachment and mobile banking service quality with positive word-of-mouth. J. Prod. Brand Manag. **30**(8), 1162–1175 (2021). https://doi.org/10.1108/JPBM-02-2020-2747

Ram, S., Sheth, J.N.: Consumer resistance to innovations: the marketing problem and its solutions. J. Consum. Mark. **6**(2), 5–14 (1989). https://doi.org/10.1108/EUM0000000002542

Rathore, B.: Exploring the impact of digital transformation on marketing management strategies. Eduzone: Int. Peer Rev./Refer. Multidisc. J. **8**(2), 39–48 (2019)

Reis, J., Melão, N.: The path to digital transformation: overcoming prejudice in the digital era with service operations. Int. J. Serv. Oper. Manag. (IJSOM) **39**(1), 81–97 (2021)

Rogers, E.M.: Diffusion of Innovations, 4th edn. The Free Press, New York (1995)

Roh, T., Park, B.I., Xiao, S.S.: Adoption of AI-enabled Robo-advisors in Fintech: simultaneous employment of UTAUT and the theory of reasoned action. J. Electron. Comm. Res. **24**(1), 29–47 (2023)

Roy, S.K., Balaji, M.S., Quazi, A.M., Quaddus, M.: Predictors of customer acceptance of and resistance to smart technologies in the retail sector. J. Retail. Consum. Serv. **42**, 147–160 (2018)

Rust, R.T., Kim, J., Dong, Y., Kim, T.J., Lee, S.: Drivers of customer equity. In: Handbook of Research on Customer Equity in Marketing, pp. 17–43. Edward Elgar Publishing (2015)

Rust, R.T., Lemon, K.N., Zeithaml, V.A.: Return on marketing: using customer equity to focus marketing strategy. J. Mark. **68**(1), 109–127 (2004)

Nouri, S.: Diversity and Inclusion in AI, Forbes Technology Council (2021). Accessed 16 Mar 2021

Samanta, I., Niros, M.I., Pollalis, Y.A.: Effects of market orientation, strategies and actions on SMEs' performance during fragile economic conditions. Arch. Bus. Rev. **8**(6), 96–117 (2020)

Sanz-Blas, S., Buzova, D., Miquel-Romero, M.J.: From Instagram overuse to instastress and emotional fatigue: the mediation of addiction. Span. Jo. Mark. - ESIC **123**(2), 143–161 (2019). https://doi.org/10.1108/SJME-12-2018-0059

Sestino, A., Prete, M.I., Piper, L., Guido, G.: Internet of Things and Big Data as enablers for business digitalization strategies. Technovation **98**, 102173 (2020)

Shu, Q., Tu, Q., Wang, K.: The impact of computer self-efficacy and technology dependence on computer-related technostress: a social cognitive theory perspective. Int. Jo. Hum.-Comput. Interact. **27**, 923–939 (2011). https://doi.org/10.1080/10447318.2011.555313

Silberstein, N., Shoham, O., Klein, A.: Combating ad fatigue via frequency- recency features in online advertising systems. In: Proceedings of the 32nd ACM International Conference on Information and Knowledge Management (CIKM '23), pp. 4822–4828. Association for Computing Machinery, New York (2023). https://doi.org/10.1145/3583780.3615461

Spita, M., Peitzika, E., Chatzi, S.: Social media adoption among small and medium-sized Greek hotels: a survey about its antecedents and its impact on performance outcomes. Int. J. Decis. Sci. Risk Manag. **9**(1–2), 23–54 (2020)

Srinivasan, R., Sarial-Abi, G.: When algorithms fail: consumers' responses to brand harm crises caused by algorithm errors. J. Mark. **85**(5), 74–91 (2021). https://doi.org/10.1177/002224292 1997082

Strohbach, M., Daubert, J., Ravkin, H., Lischka, M.: Big data storage. In: Cavanillas, M., Curry, E., Wahlster, W. (eds.): New Horizons a Data-driven Econ, pp. 119–141. Springer, Heidelberg (2016)

Susser, D., Roessler, B., Nissenbaum, H.F.: Technology, autonomy, and manipulation. Internet Policy Rev. **8**(2) (2019). https://doi.org/10.14763/2019.2.1410. SSRN: https://ssrn.com/abs tract=3420747

Talwar, S., Talwar, M., Kaur, P., Dhir, A.: Consumers' resistance to digital innovations: a systematic review and framework development. Aust. Mark. J. **28**(4), 286–299 (2020)

Taylor, C.R.: Ad skipping, the 'ad free internet' and privacy: a call for research. Int. J. Advert. **43**(3), 403–404 (2024). https://doi.org/10.1080/02650487.2024.2309747

Teng, X., Wu, Z., Yang, F.: Research on the relationship between digital transformation and performance of SMEs. Sustainability **14**(10), 6012 (2022)

Thompson, D.V., Hamilton, R.W., Rust, R.T.: Feature fatigue: when product capabilities become too much of a good thing. J. Mark. Res. **42**(4), 431–442 (2005). https://doi.org/10.1509/jmkr. 2005.42.4.431

Tsamados, A., et al.: The ethics of algorithms: key problems and solutions. AI Soc. **37** (2021). SSRN: https://ssrn.com/abstract=3662302. https://doi.org/10.2139/ssrn.3662302

Tzoumaka, E., Leivadi, S., Kaplanidou, K.: Sport tourism in the VUCA era: the effect of destination image on behavioral intentions of active sport tourists. In: Zarkada, A. (ed.) Marketing Solutions to the Challenges of a VUCA Environment. GMA-GAMMA 2023. Springer Proceedings in Business and Economics. Springer, Cham (2024). https://doi.org/10.1007/978-3-031-58429-9_4

Ukanwa, K., Rust, R.T.: Algorithmic bias in service. USC Marshall School of Business Research Paper (2021). SSRN: https://ssrn.com/abstract=3654943. https://doi.org/10.2139/ssrn.3654943

Verma, S., Sharma, R., Deb, S., Maitra, D.: Artificial intelligence in marketing: Systematic review and future research direction. Int. J. Inf. Manag. Data Insights **1**(1), 100002 (2021)

Vermeer, S.A.M., Araujo, T., Bernritter, S.F., van Noort, G.: Seeing the wood for the trees: how machine learning can help firms in identifying relevant electronic word-of-mouth in social media. Int. J. Res. Mark. **36**(3), 492–508 (2019)

Wieseke, J., Geigenmüller, A., Kraus, F.: On the role of empathy in customer-employee interactions. J. Serv. Res. **15**(3), 316–331 (2012). https://doi.org/10.1177/1094670512439743

World Economic Forum. Small Business, Big Problem: New Report Says 67% of SMEs Worldwide Are Fighting for Survival (2022a). Accessed 21 Sept 2024. https://www.weforum.org/press/2022/

World Economic Forum. Why we shouldn't overlook the impact of SMEs on local and global economies (2022b). Accessed 21 Sept 2024. https://www.weforum.org/agenda/2022/

Yam, K.C., et al.: Robots at work: People prefer—and forgive—service robots with perceived feelings. J. Appl. Psychol. **106**(10), 1557–1572 (2021). https://doi.org/10.1037/apl0000834

Yu, C., Chantatub, W.: Consumers' resistance to using mobile banking: evidence form Thailand and Taiwan. Int. J. Electron. Comm. Stud. **7**, 21–38 (2016)

Zaki, H.O., Fernandez, D., Dastane, O., Aman, A., Sanusi, S.: Virtual reality in digital marketing: research agenda based on bibliometric reflection. Mark. Intell. Plan. **41**(4), 505–524 (2023)

Zarkada, Anna K.: Marketing solutions to the challenges of a vuca environment: what have we accomplished and what more can we do? In: Zarkada, A. (ed.) Marketing Solutions to the Challenges of a VUCA Environment: Proceedings of the 2023 GMA-GAMMA Joint Symposium, Seoul, Korea, July 20-23, 2023, pp. 1–14. Springer, Cham (2024). https://doi.org/10.1007/978-3-031-58429-9_1

Zennyo, Y.: Freemium competition among ad-sponsored platforms. Inf. Econ. Policy **50**, 100848 (2020). https://doi.org/10.1016/j.infoecopol.2019.100848

Zheng, H., Ling, R.: Drivers of social media fatigue: a systematic review. Telemat. Inf. **64**, 101696 (2021)

Ziółkowska, M.J.: Digital transformation and marketing activities in small and medium-sized enterprises. Sustainability **13**(5), 2512 (2021)

Zuboff, S.: Big other: surveillance capitalism and the prospects of an information civilization. J. Inf. Technol. **30**(1), 75–89 (2015). https://doi.org/10.1057/jit.2015.5

SMEs Playing in the Big Advertising League: The Transformative Potential of Programmatic Advertising and the Barriers to Its Full-Scale Implementation

Marius Ian Yul Horstick[1] and Anna Zarkada[2(✉)]

[1] MHS Media Solutions, Karlheinz-Steimel-Weg 15, 50969 Cologne, Germany
horstick@mhs-media.com
[2] Department of Communication and Marketing, Cyprus University of Technology 30, Archbishop Kyprianou Street, 3036 Limassol, Cyprus
anna.zarkada@cut.ac.cy

Abstract. Programmatic advertising has emerged as a transformative digital marketing tool offering precise audience targeting, scalability, efficiency, and resource optimization through advanced technologies such individual targeting options, optimized display-ad distribution, and real-time bidding. While large corporations have widely adopted programmatic advertising, SMEs face challenges such as technical complexity, high costs, lack of expertise, and GDPR compliance. Programmatic advertising offers SMEs opportunities to optimize budget allocation and enhance advertising efficiency, yet adoption barriers including transparency concerns, infrastructure demands, and workforce limitations persist.

Drawing from a synthesis of academic literature, industry reports, and direct insights from industry professionals, including marketing and advertising decision-makers from SMEs as well as high-level agency executives, we highlight both the benefits and challenges of PA adoption for European SMEs.

A Four-Phase Framework is proposed to guide SMEs through analysis, planning, execution, and evaluation of programmatic advertising, emphasizing gradual adoption, investment on technical infrastructure, and workforce training. Overcoming operational and regulatory hurdles is essential to unlocking PA's full benefits and ensuring sustainable competitiveness in the evolving digital advertising landscape.

Keywords: Programmatic Advertising · GDPR Compliance · Resource Constraints · Technological Infrastructure · Germany · SMEs

1 Introduction

Programmatic advertising (PA), the automated, data-driven ad buying and placement, has fundamentally reshaped digital marketing; so much so that it has been described as a cornerstone of contemporary marketing strategies [1]. Indeed, since 2023 PA constitutes over half of all non-social display advertising expenditures, with a total value

A. K. Zarkada (Ed.): GMA-GAMMA 2025, SPBE, pp. 205–239, 2025.
https://doi.org/10.1007/978-3-031-96930-0_11

of €96.6 billion [2] and has expanded beyond desktop platforms to mobile, video, and other digital channels [3, 4]. Its success rests on its ability to process advertising transactions within milliseconds, enabling advertisers to adapt swiftly to evolving consumer preferences and behaviors. This rapid functionality is supported by centralized data systems and algorithms capable of analyzing vast amounts of digital signals, including user demographics, interests, and online behaviors, to identify and engage relevant audiences [5–8].

By seamlessly integrating advertisers, publishers, and advanced technologies, including artificial intelligence (AI), machine learning, and Real-Time Bidding (RTB), PA facilitates the streamlined interaction between the buy-side and sell-side of digital advertising transactions [9, 10]. Moreover, by dynamically adjusting ad placement based on user behavior and contextual factors it efficiently delivers precisely tailored messages to specifically targeted audiences at optimal times thus achieving superior campaign scalability and performance, optimized resource allocation, and maximized return on investment (ROI) [11–18].

The transformative potential of PA has been recognized for years now [3] but in academia, this constellation of technologies is still under-researched and disassociated with practice [19]. Moreover, what little the academic literature has to offer about PA [20, 21] is based on studies of large multinational corporations and US-based firms, thus dominated by industries reports and handbooks. In Europe, and in Germany in particular, however, the backbone of the economy is small and medium sized enterprises (SMEs) [22] the specific challenges of which are notably ignored in the PA literature [11–13, 21, 23, 24].

This chapter synthesizes insights from various sources academic, industry, anecdotal and experiential (Discussions and direct insights from marketing and advertising decision-makers in German SMEs, in the B2B and retail sectors as well as senior agency executives and account managers) into an outline and explanation of PA which is accessible to the average marketer and business manager or owner of an SME. As we both live, teach, and work in Europe, we obviously see things from a predominantly western perspective and an understanding of SMEs as relatively simple in structure, usually family or owner-managed companies with less than 250 employees. We place an emphasis on the German context not only for convenience but also because Germany is the largest EU economy but still typical of the structure of the whole common market which is dominated by small by international standards SMEs.

1.1 A Brief Note on the German Context

The introduction of PA into the German market has evolved over time. First movers included the Trade Desk, a California-based company, which began operations in 2013 and Gruner + Jahr's 2016 programmatic booking initiative [25]. PA has since become central to digital marketing in Germany, projected to account for nearly three quarters of advertising revenue by 2024, totaling approximately €4.4 billion [15, 26, 27]. The internet now leads German advertising sales, surpassing traditional media [28].

Germany's PA ecosystem is supported by institutions like the Interactive Advertising Bureau (IAB) and Bundesverband Digitale Wirtschaft (BVDW), which set standards and provide training. Key players include technology providers such as Microsoft, Adform, and Google's DV360, publishers like Improve Digital and Magnite, and service providers like Tradedoubler and Xaxis. Data management platforms (DMPs) like eXelate and Microsoft Invest further enhance PA by leveraging first party data [29, 30]. Events like d3con drive collaboration, while growing digital competition underscores the need for German SMEs to engage in PA [31–34].

1.2 Aim and Objectives

The aim of this chapter is to explore PA in the context of European SMEs. The objectives are:

(a) To demystify PA by providing a structured and comprehensive overview of the technology with an emphasis on the components of the PA ecosystem and their operational roles.
(b) To assist the non-expert marketer in evaluating the relative merits of alternative buying methods, operational processes, with an emphasis on targeting options, and PA effectiveness measures by outlining simple and easily applicable criteria for selecting between alternatives and evaluating decision outcomes.
(c) To provide a practical framework and SME-targeted recommendations for adopting PA.

The chapter is organized in the logic of the objectives. First, it describes the elements of the PA ecosystem and how they work together, it explains the buying and operational process and presents the commonly used PA effectiveness metrics. Then it discusses the benefits, challenges, and barriers to PA implementation by SMEs. Finally, a four-phase framework for PA implementation is presented and analyzed.

2 The Programmatic Advertising Ecosystem

The PA ecosystem is a sophisticated, data-driven framework that automates the buying and selling of digital advertisements. This system revolves around two primary stakeholders: advertisers on the buy-side and publishers on the sell-side. These parties interact through a set of interdependent platforms and technologies designed to optimize the efficiency and effectiveness of ad transactions. Central to this ecosystem are Supply-Side Platforms (SSPs), Demand-Side Platforms (DSPs), Data Management Platforms (DMPs), Ad Servers, and Ad Exchanges, each playing a distinct role in facilitating the delivery of targeted advertisements [5–7]. These we briefly present.

2.1 Supply-Side Platforms

Supply-Side Platforms (SSPs) are integral to the sell-side of PA, enabling publishers to aggregate, manage, and sell advertising inventory to a wide array of potential buyers (Learn Ad Operations, 2024). By connecting publishers with multiple advertisers, SSPs function as central hubs for managing ad inventory across various channels, including websites, mobile applications, and video platforms. SSPs employ programmatic technologies such as RTB to maximize revenue by allowing advertisers to compete for available ad space through automated auctions. This competitive process ensures publishers receive the highest possible price for their inventory.

Beyond revenue optimization, SSPs provide publishers with tools to control pricing, manage inventory access, and customize ad placement visibility. These capabilities empower publishers to maintain granular control over their inventory, ensuring alignment with branding and content strategies [6, 7, 35, 36]. Additionally, the centralized functionality of SSPs enhances operational efficiency by streamlining the traditionally labor-intensive processes of inventory management and ad sales. As a result, SSPs play a critical role in helping publishers maximize the value of their inventory while retaining control over the advertising ecosystem.

2.2 Demand-Side Platforms

On the buy-side, Demand-Side Platforms (DSPs) provide advertisers with tools to manage, purchase, and optimize digital advertising campaigns in real-time. These platforms aggregate ad inventory from multiple publishers and exchanges, offering advertisers a unified interface to define campaign parameters, allocate budgets, and establish targeting criteria. By automating the ad-buying process, DSPs enable advertisers to focus on strategic decision-making while the platform executes bidding and ad placements [12, 13, 37].

A key strength of DSPs lies in their ability to leverage data to refine ad targeting. By incorporating insights from audience demographics, behavioral patterns, location, and contextual data, DSPs allow advertisers to bid on impressions that align closely with their campaign objectives [7, 12, 13, 37]. Furthermore, real-time analytics provided by DSPs enable advertisers to monitor campaign performance and make data-driven adjustments, enhancing both efficiency and cost-effectiveness. Through advanced targeting algorithms and automated bidding processes, DSPs empower advertisers to maximize ROI while reaching highly relevant audiences with precision [38].

It is obvious that the key ingredient of the marvelous PA gateaux is data. Where it comes from, how it is stored and managed, who protects it and how, we explore in the following section.

2.3 Data Management Platforms

Data Management Platforms (DMPs) form the backbone of the PA ecosystem as they facilitate the collection, organization, and analysis of vast amounts of consumer data from various sources. They are centralized systems designed to perform accurate segmentation and customer profiling to facilitate targeting, and personalization of digital advertising and marketing campaigns. To achieve these, they often work in conjunction with DSPs and customer relationship management (CRM) systems [7, 12, 13].

DMPs function in four key stages:

1) Data collection: DMPs aggregate data from multiple sources such as website interactions, corporate social media accounts but also scraping of data from user and

competitor accounts, online ads performance metrics such as click-through rates, and CRM databases which include data related to online and offline purchases, customer complaints etc. The data they collect can be both structured (such as databases) and unstructured (such as text and images) and as diverse as demographics, biometrics collected from smart wearables devices, purchase history, devices used to access online content, and how often people vacuum or mop their floors using data from their robotic vacuum cleaners.

2) Data cleaning, organization and processing: The platforms classify data according to predetermined rules and impose structural patterns on the data. DMPs currently in operation have high degrees of autonomy and unprecedented, and ever-increasing, processing speed and ability to handle complex operations as the utilize artificial intelligence (AI) and machine learning. Thus, they can not only combine multiple data sets and identify patterns even in the most unstructured ones, but they can also predict customer behaviors with high levels of accuracy through data transformation, feature selection, statistical pattern evaluation and interpretation. DMPs can perform various data mining functions such as *inter alia*:
   a. Supervised learning: categorizing cases on the bases of already labelled data (e.g. male/female customers).
   b. Unsupervised learning: detecting underlying patters in unlabeled data (e.g. behavioral and values, attitudes, lifestyle (VALS) segmentation).
   c. Association Rules: discovering relationships in data as in people who buy a mouse are also buying mouse pads.
   d. Neural networks which simulate human information processing much like our brains.
3) Audience segmentation and insights: The data are contextualized and integrated into custom audience segments which are then used to design and execute personalized marketing and advertising campaigns.
4) Data activation: DMPs integrate with DSPs, SSPs, to execute PAcampaigns.

Examples of DMPs are Adobe Audience Manager™, Oracle BlueKai™, Nielsen Marketing Cloud™, and Google Audience Center™.

Data Sources

DMPs integrate data from multiple sources such as user interactions with a company's website or apps, customer interactions with partner organizations, and aggregated consumer data collected by external providers and offered for sale [6, 7, 12, 13, 39, 40]. Further to their source, the key differentiators between types of data are their availability and quality. These two characteristics are in an inverse relationship (diagrammatically presented in Figure 1 and discussed in detail below.

The three categories of data DMPs use, and the main considerations of their use are explained herewith:

(a) First-Party Data are collected directly by advertisers or publishers from their own audiences, often through their in-house CRM platforms. They include user information such as website visits, demographic details, purchase history, and behavioral patterns. First-party data is highly valued for its accuracy and relevance, as it directly reflects company – user real interactions. Their collection, storage, management and

Fig. 1. Schematic categorization of types of data used in PA. Source: authors' own work based on O'Hara, 2016 [40]

utilization, however, can be resource-intensive, requiring substantial investments to ensure data quality and security. Despite the costs, first-party data remains integral to precise ad targeting, fostering stronger connections between advertisers and their audiences [7, 13].

(b) Second-Party Data are obtained through partnerships between companies allowing access the each other's first-party data through DMPs. This allows businesses to enhance their targeting capabilities by leveraging insights from trusted partners, thus reducing reliance on third-party data providers, broadening audience reach, and improving ad placement accuracy whilst mitigating some of the privacy and quality concerns associated with third-party data sources [7, 13].

(c) Third-Party Data are collected by data brokers, specialized organizations who are external to the advertisers, and made available for purchase. They are broad datasets compiled from multiple sources which means that they are often of lower quality compared to firstor second-party data. Moreover, because of the increasing enforcement of data protection regulations, such as the General Data Protection Regulation (GDPR), third-party data carry additional privacy and legal risks, as well as significant ethical concerns [8, 39] which means that their use must be carefully managed.

The data pool, to which advertisers and agencies have access, enables them to create detailed customer and prospects profiles on which they base their targeting strategies [7, 12, 13, 37]. Thus, DMPs play a pivotal role in audience segmentation and modelling. These processes help advertisers identify and engage existing and potential customers which share common or similar characteristics. By analyzing user behavior, purchase history, and contextual factors, DMPs generate actionable insights that support more effective campaign decision-making [7, 41]. These insights are seamlessly integrated into DSPs, enhancing targeting parameters and improving campaign performance.

The above discussion makes accessing and utilizing data sources look like a straightforward process handled by tech giants on behalf of companies who reap the benefits of

21st Century consumers' addiction to selfies, stories, and indiscreetly sharing what they eat, buy, think, hate and love for the world to see [42–44]. Alas! It is not so simple for SMEs.

All that Data is Not Gold

There is a good reason why for decades business intelligence and, in particular, marketing data collection, management, and utilization using statistical, machine learning, and database management techniques has been described as 'data mining': You need to dig wide and deep, shift through ore, overburden, and tailings – that is garbage to the uninitiated – and process tons of materials to get a nugget of gold. Much as data is the cornerstone of PA, driving audience targeting, segmentation, and campaign optimization, a number of obstacles stand in the way of effectively leveraging data to access the full potential of the PA ecosystem. The smaller the organization, the larger the obstacles.

SMEs have relatively limited access to high-quality first-party datasets as their CRM systems are rarely fully deployed and automated. They usually participate in, small networks of associated businesses, and even when they have access to networks involving large and structured partners, they only have low levels of power within these networks and so find it hard to acquire second-party data. Good quality, reliable, and accurate third-party data is costly and so often beyond the means of SMEs with limited marketing budgets. Much like the old jokes about dogs chasing cars, even if an SMEs, somehow, gains access to a good data set, knowing what to do with it and how best to exploit it requires infrastructure which is costly and requires high levels of expertise to develop and utilize. For European SMEs, compliance with the strict EU data protection regulations, such as GDPR, adds further complexity and imposes the need for dedicated data management resources [9, 15, 17, 45–51]. The data-related challenges SMEs face in implementing PA programs are summarized in Table 1 alongside their description and impact on marketing operations.

The last of the data-related challenges which European SMEs face when using DMPs for PA is particularly important and requires further elaboration.

The Specifics of General Data Protection Regulation

Regulatory issues and data privacy concerns have ethical, legal, operational, marketing, and financial implications. Strict adherence to regulations which aim to ensure ethical handling of consumer data and safeguard the public's right to data privacy, such as GDPR, is an imperative of corporate citizenship [52–55] but also a barrier to PA implementation for SMEs, especially in the EU. Given the sensitivity of personal information, compliance with data management regulations is essential for maintaining consumer trust and avoiding legal consequences [39] which can have wide reaching and hard to bear financial and reputational consequences.

The most important, in terms of legal and operational burdens placed on businesses in the EU [56, 57], regulatory framework affecting PA is GDPR, a summary of the principles of which (based on Article 5.1 which outlines seven key data privacy principles [58] , is presented here:

1) Lawfulness, fairness, and transparency: To ensure legal compliance, it is crucial to identify valid grounds for processing personal data. The Common Information Security Control outlines six justifications:

Table 1. Data-related challenges of PA implementation by SMEs

Challenge	Description	Impact on Businesses
Access to data	Due to underdeveloped CRM, limited access to networks of associated businesses, and budget constraints, SMEs struggle to acquire extensive high-quality datasets	Inaccurate segmentation. Incorrectly targeted audiences. Ineffective personalization Insufficient campaign optimization potential
Data acquisition costs	Acquiring quality data from reliable sources is costly, often beyond the financial scope of an SME marketing budget	
Infrastructure	Robust data management requires high initial investment in technical infrastructure, high maintenance and running costs, and access to costly and often hard to find expertise	Inadequate data storage, management and analysis lead to poor or ineffective decision-support, resource wastage, and suboptimal ad placement Poor data storage and handling incurs legal penalties and added compliance and litigation costs
Compliance with evolving data privacy regulations	There is a marked shift away from third-party cookies which complicates ad personalization and restricts and strictly regulates retargeting	SMEs need to invest in costlier alternatives like first-party data and contextual targeting Data-compliant, privacy-friendly solutions require redeployment of limited resources and strategic realignment of marketing approaches to maintain both relevance and compliance

a. consent,
b. contract,
c. legal obligation,
d. vital interests,
e. public task, and
f. legitimate interests.

At least one must be present to meet GDPR requirements for lawful data processing.

2) Purpose limitations: Following this concept ensures data subjects understand why their data is collected and have reasonable expectations about its management. It also gives them control over future use and the choice to provide or withhold their information.

3) Data minimization: GDPR mandates companies to collect only the minimum data necessary for a specific purpose. It emphasizes relevance and necessity, prohibiting data collection without a clear purpose, though it allows data collection for known future needs.
4) Accuracy: Every reasonable step must be taken to ensure that personal data that are inaccurate, or out-of-date, or not strictly conforming to the purposes for which they are processed, are erased, or rectified without delay.
5) Storage limitation: GDPR mandates personal data to be kept only if necessary for the intended purpose. Companies must justify their retention periods as longer durations increase the risk of data becoming inaccurate or outdated.
6) Integrity and confidentiality: This principle covers both internet and physical security. GDPR mandates that only authorized personnel can access personal data and requires recovery measures in case data is leaked to unauthorized parties, lost, altered, or damaged.
7) Accountability: This GDPR requirement emphasizes personal responsibility in handling data and adherence to set criteria. Records and measures must demonstrate compliance, and in case of a breach, safeguards should show efforts to prevent or rectify it.

As GDPR requires explicit user consent for data processing and so it fundamentally alters earlier digital business models, such as the free-for-all digital marketing landscape of the 2000s, and fundamentally transforms PA into a complex and strictly delineated process. It mandates that businesses implement measures like privacy notices, consent management platforms (CMPs), and data anonymization [59, 60]. Implementing its principles, such as the principle of data minimization, complicate the process of finding the balance between effective advertising and fulfilling privacy obligations [61]. Frameworks like the IAB´s 'Europe Transparency and Consent Framework' offer guidance but face limitations and legal challenges, adding uncertainty [62, 63]. Non-compliance risks penalties of up to 4% of global annual turnover, further pressuring companies to adapt PA [64].

Additionally, GDPR has accelerated the decline of third-party cookies, which had been the bread and butter of PA, forcing businesses to invest in first-party data strategies to maintain competitive capabilities [10, 65]. These efforts are essential for aligning PA practices with ethical and legal standards, but they demand significant investment and a focus on building and systematically maintaining user trust through transparency and explicit consent [39, 66]. The expansion of data-driven advertising practices is unlikely to be halted or reversed. It seems that its advance, however, will place ever more pressure on companies to strengthen and expand their legal resources to be able to effectively manage compliance in areas like user consent and data processing [67, 68]. For the German small marketing agencies, we interviewed the regulatory environment makes programmatic advertising difficult to implement due to the high initial costs setting up a compliance system requires. One of them characteristically noted that if there ever is an end to cookies from third-party providers, PA in its current form will naturally also come to an end.

It is clear from the above discussion of data-related issues that balancing datadriven PA precision with ethical and regulatory considerations is part and parcel of the DMP function. Employing the services of DMPs which are not only reliable but also fully compliant with EU regulations enables advertisers to achieve their objectives whilst fending off the moral, legal, financial, and reputational risks incurred by being caught, or being perceived to be, in violation of regulatory standards. The ethical and safe approach to PA necessitates costly investments in compliance measures, IT systems, tech personnel salaries, and lawyers' fees, which often sum up to way beyond what an SME can comfortably afford. Challenging as responsible data management and compliance with privacy regulations might prove for advertisers, they, nevertheless, promote transparency and increase trust in the advertisers and the advertising industry. This way they ultimately benefit the digital advertising ecosystem and its users in the long term [69].

2.4 Ad Servers and Ad Exchanges

Ad Servers and Ad Exchanges (AdX) comprise the technical and transactional infrastructure of PA. Ad servers manage the storage, delivery, and tracking of advertisements, ensuring they are displayed to the intended audience. These platforms collect performance metrics such as impressions, clicks, and conversions, which are crucial for evaluating campaign effectiveness and implementing data-driven optimizations. The data gathered by ad servers often integrates with DMPs, enhancing the accuracy of targeting strategies [7, 12, 13, 37].

In contrast, ad exchanges operate as digital marketplaces where SSPs and DSPs facilitate the real-time buying and selling of ad inventory. These exchanges enable publishers to offer their inventory to a broad network of potential buyers, while advertisers can dynamically bid on impressions that meet their targeting criteria. The integration of ad servers and DMPs within ad exchanges support seamless functionality, ensuring the scalability and precision of programmatic transactions [7]. Ad exchanges exemplify the collaborative nature of the PA ecosystem, bridging the gap between buy-side and sell-side stakeholders through sophisticated technology and automated processes.

All the components of the PA ecosystem (diagrammatically presented in Fig. 2) collectively support the design and implementation of efficient data-driven advertising strategies. Figure 2 illustrates the technical process and the relationships between advertisers (buy-side) and publishers (sell-side) as well as the roles of SSPs, DSPs, DMPs and AdXs. It is important to remember that the PA ecosystem is a highly interconnected network of platforms and technologies designed to simultaneously optimize the delivery and effectiveness of digital advertisements. Each component plays a distinct role in streamlining the ad-buying process, from inventory management and data analysis to real-time bidding and campaign optimization but they can only function when they are fully coordinated in real-time. SSPs enable publishers to maximize revenue and control inventory, while DSPs provide advertisers with the tools to target audiences with precision. DMPs underpin data-driven decision-making by integrating and analyzing consumer data, and ad servers and AdXs ensure the seamless execution of transactions and ad delivery.

The PA ecosystem is a dynamic and scalable system that empowers advertisers and publishers to achieve their goals in an increasingly complex digital landscape. Understanding these elements enables advertisers to make informed decisions, optimize reach,

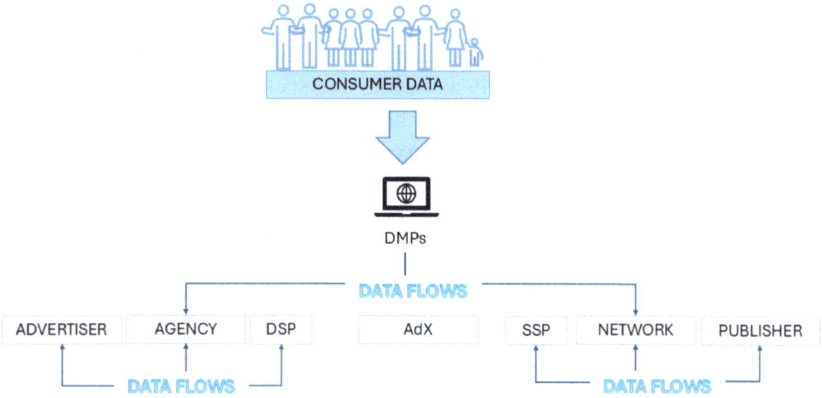

Fig. 2. Diagrammatic representation of the PA ecosystem. Source: authors' own work based on Bakare and Tolulope (2021) [70].

and ensure strategic placements while adhering to data privacy standards. Moreover, they need to be familiar with the technical process and buying methods of PA.

Building on existing insights [10, 13], highlights that the "black box" nature of programmatic advertising (PA) systems is exacerbated by staffing limitations. Interviewed business owners, CMOs, and online marketing managers from agencies and retail companies consistently cited a lack of technical expertise as a major challenge. Many described PA systems as opaque, leading to limited trust, particularly at the management level. Transparency concerns, coupled with the complexity of these systems, make internal operation difficult without specialized IT personnel. Moreover, many companies lack the technical maturity to successfully implement PA, often realizing too late that their internal capabilities are insufficient. This results in uncertainty regarding technical requirements and skepticism toward service providers and agencies.

3 Programmatic Advertising Buying Methods

Within the PA ecosystem, several technical processes govern the buying and selling of ad inventory, enabling the efficient execution of campaigns and optimizing ad placements to meet specific objectives. The key mechanisms of PA and their distinct advantages and tailoring potential to advertisers' and publishers' objectives (based on [11–13, 37] are presented here.

RTB is the prevalent form of PA. It enables advertisers to bid for ad impressions in real-time as users interact with websites. This automated auction system facilitates competition for ad placements, with the highest bidder securing the ad spot. Its scalability makes it highly effective for advertisers seeking to dynamically adjust bidding strategies based on current market demand and audience relevance. Transactions in RTB are facilitated through ad exchanges, which connect Demand-Side Platforms (DSPs) with Supply-Side Platforms (SSPs). Data Management Platforms (DMPs) and AdXs provide critical data and tracking support to enhance targeting precision as shown in Figure 2 [7, 11–13, 37]. RTB utilizes two primary auction models:

a) First-Price Auction during which the highest bidder pays the full amount of their bid, securing competitive, high-value placements. This model necessitates strategic bidding to manage costs effectively and avoid overspending.
b) Second-Price Auction, during which the highest bidder pays slightly more than the second-highest bid, fostering cost-effective bidding and encouraging broader market participation by reducing the risk of overpaying.

Both these auction models provide businesses with flexible, cost-efficient options for securing ad placements [7, 11–13, 37]. The PA ecosystem relies on both auction types and technical inventory purchase flow processes resulting in different transaction types. These processes coordinate the buying and selling mechanisms in specific configurations which align campaign objectives with the desired degree of control over ad placement and the available budget.

Purchasing ad inventory can take place using different auction or contractual arrangements, each offering varying levels of exclusivity, control, and pricing stability. As with all advertising decisions, the choice of PA transaction type is a function of campaign goals, budget, and brand safety requirement. The four primary methods are described and discussed based on a synthesis of academic and commercial and industry sources [7, 11–13, 17, 37, 71–73]:

(a) Open Auctions are the most accessible buying method within the programmatic ecosystem and the prevalent one due to its scalability and costefficiency. Publishers list their inventory on SSPs, allowing all connected advertisers to compete for ad impressions. Pricing is determined through competitive bidding, making this method highly scalable for advertisers seeking broad reach. Although Open Auctions offer limited control over placement, advertisers can employ blacklists to exclude undesired websites or environments, thereby ensuring a degree of brand safety.
(b) Private Auctions, also known as invitation-only auctions, provide a more controlled and selective approach to inventory sales. Publishers restrict access to a curated group of advertisers, offering security, exclusivity and enhanced control over ad placements. Conducted within Private Marketplaces (PMPs), these auctions prioritize brand safety and premium inventory, minimizing the risk of association with unsuitable content.
(c) Preferred Deals, also known as unreserved fixed rate, are non-auction-based arrangements where advertisers and publishers negotiate a fixed price for inventory. While inventory availability is not guaranteed, this method offers pricing stability and eliminates the uncertainties associated with fluctuating auction prices. Preferred Deals strike a balance between flexibility and predictability, making them suitable for advertisers who require cost consistency without fully committing to guaranteed inventory.
(d) Automated Guaranteed or programmatic direct, represents the most structured method of programmatic ad buying. In this model, advertisers and publishers directly negotiate both fixed pricing and guaranteed inventory delivery. Unlike auction-based methods, Automated Guaranteed automates the transaction process while ensuring access to premium inventory and guaranteed ad impressions. This approach is particularly advantageous for advertisers seeking high-value placements in prominent

digital environments, as it combines efficiency with reliability in delivery and brand visibility.

The PA inventory transactions follow a distinct sequence across Open Auctions, Private Auctions, Unreserved Fixed Rates, and Automated Guaranteed methods. The process begins with inventory made available by publishers on SSPs, where it is either open to all advertisers or restricted to a select few, depending on the chosen method. AdX and Ad-Servers facilitate these transactions by connecting DSPs and SSPs in real time, ensuring that each ad reaches its target audience effectively. These methods enable advertisers and publishers to optimize their approach based on campaign needs, choosing between highly competitive, flexible bidding options and more controlled, premium placements.

Each transaction method and data type in PA supports flexible options, such as Open Auctions for dynamic bidding or Programmatic Direct for guaranteed placements. Businesses can align these methods with their campaign goals to balance efficiency and control with data privacy and regulatory compliance restrictions and resource constraints, including the high costs of access to high-quality datasets [9, 45].

4 Programmatic Advertising Operational Process

PA follows the logic of traditional media planning and booking albeit transformed into a dynamic real-time process through automation and, data-driven decisionmaking [74]. Understanding this transformation process is crucial for advertisers and agencies both to fully harness the potential of PA to minimize ad spend waste and to accurately target predefined audiences [6, 7]. The PA ecosystem components such as data, targeting options, and real-time auctions are fully integrated but to make the process accessible to a general business and marketing audience we provide a breakdown illustrating how each element contributes to maximizing ad relevance and optimizing campaign resource use.

When a user lands on a webpage the publisher's ad server sends a signal to the SSP, indicating ad inventory availability. Advertisers, through DSPs, access this information and evaluate the available inventory based on audience profiles and bidding strategies that are optimized with first-, second-, and third-party data. This process triggers RTB, where the advertiser with the highest bid wins the ad placement. The banner for the winning ad is then transmitted to the publisher's ad server and displayed within milliseconds, ensuring that ads are bought and displayed at remarkable speed. This process is repeated with each page load, allowing advertisers to continuously optimize their campaigns based on user data and campaign performance [6, 7, 35, 75].

PA accommodates all the targeting methods used in traditional advertising but also introduces some new technology-based ones that ensure unprecedented levels of precision and relevance in ad delivery. The most common targeting options available within PA [7, 76–82] are presented in Table 2 which explains each targeting method and gives examples of their application possibilities.

Table 2. Targeting methods commonly used in PA

Targeting Method	Description	Example
Demographic	Advertising based on demographic characteristics like age, gender, income, etc	Targeting an audience aged 1835 with middle income
Geographic	Ads targeted by location, such as country, city, or region	Advertising for a store to users found within a 5 km radius
Behavioral	Advertising based on users' online behavior, such as visited sites or search queries	Targeting users who recently searched for travel to China
Contextual	Ads appear on websites relevant to the advertising message	A fitness ad on a sports website
Device	Advertising tailored to specific device types	Mobile-optimized ads displayed only on smartphones
Lookalike	Targeting audiences like existing customers	A face cream which is usually bought by middle-aged women is promoted to middle-aged women who have never bought it before
Re-Targeting (Remarketing)	Ads for users who have already interacted with a website or ad	Targeting a user who abandoned the shopping cart
Contextual Keyword	Ads appear on pages with specific keywords relevant to the target audience	Ads for travel insurance on pages with terms like 'vacation' or 'travel'
Interest and Life Stage	Ads based on users' interests or current life stages	Targeting users who recently became parents
Cross-Device	Targeting users across different websites and devices, understanding when a user is likely to switch between devices	Showing consistent ads to a user across their smartphone and laptop
Technical	Targeting based on technical as- pects like operating systems or device settings	Displaying specific ads for users on iOS devices
Semantic	Targeting by analyzing all visible text on a website to enhance ad relevance	Serving ads for eco-friendly products on pages where 'sustainability' is frequently mentioned

5 Programmatic Advertising Effectiveness Metrics

To assess how effectively a PA campaign attracts attention and fosters meaningful interactions it is essential to examine multiple core metrics which provide critical insights into both campaign performance and its optimization. Some of the key performance indicators (KPIs) used in PA are similar to what we use for traditional ad effectiveness measures whilst others are particular to the online environment and only available in the PA ecosystem. None of them is enough to evaluate a campaign on its own; instead, it is vital for designing efficient and cost-effective campaigns to examine a combination of KPIs. The most basic PA metrics help balance visibility with engagement to maximize results. Such KPIs are the following:

(a) Reach and Frequency are two KPIs which are usually examined together. Reach measures the number of unique users exposed to an ad, while Frequency tracks how often these users encounter the same advertisement. Together, these two metrics help advertisers balance widespread visibility with avoidance of overexposure which can lead to audience fatigue and diminished engagement [80, 83, 84].
(b) Viewability Rate measures the proportion of ads visible to users, serving as a benchmark for campaign transparency and effectiveness. In dynamic programmatic environments with diverse placements, achieving high viewability rates ensures ads are delivered as intended. Tools such as Google's Active View™ provide industry standards for optimizing viewability, thereby enhancing campaign credibility and impact [85, 86].

The purpose of an advertisement, however, is not to be viewed. The most commonly used KPIs which extend beyond basic impressions and provide nuanced insights into user behavior are the following:

(c) The most basic one is the Click-Through Rate (CTR) which measures the percentage of ad impressions that result in user clicks on the advertiser's website. This metric serves as a preliminary indicator of ad relevance and engagement. It highlights how well an ad resonates with its target audience but it is not sufficient for evaluating user intent or post-click behaviors [87]. It follows; all of us have, at some point in time, landed on a website we are not interested in by a careless mouse move. Moreover, users may click on an ad but disengage quickly, indicating limited interest or ineffective targeting. To decide which is the case we use complementary metrics, such as dwell time, bounce rates, and social interactions which offer a more comprehensive perspective by capturing post-click engagement.
(d) More nuanced than plain CTR is the Engagement Rate (ER). This metric reflects the breadth of interactions with ad content, encompassing video completions, social media shares, and interactions with dynamic elements like rich media. Particularly valuable for campaigns emphasizing brand storytelling or aimed at raising brand awareness, the engagement rate assesses both user interest and emotional connection [23, 38, 88–91]. Unlike CTR, which focuses on immediate clicks, the engagement rate metric reflects sustained and varied interactions, thus providing a more holistic understanding of user behavior.

Viewing an advertisement, visiting a website or even sharing an ad, however, does not translate to tangible marketing outcomes such as sales. To overcome this limitation CTR and ER are examined vis-a-vis outcome-based metrics which evaluate the degree to which a PA campaign drives desired user actions, such as purchases, subscriptions, or sign-ups. Metrics which bridge the gap between audience engagement and tangible business outcomes are the following:

(e) Conversion Rate (CVR) calculates the percentage of users who complete a specified action after interacting with an ad. As a direct measure of campaign effectiveness, CVR illustrates how programmatic strategies translate user interest into measurable results. For a complete interpretation of CVR analysts also take into account the segmentation criteria that were used as these significantly influence conversion behaviors [92–94].
(f) Cost Per Acquisition (CPA) measures the financial efficiency of an ad campaign by dividing total ad spend by the number of conversions achieved. This metric provides insights into the cost-effectiveness of customer acquisition efforts. For campaigns leveraging programmatic strategies, lower CPA values typically reflect effective audience targeting and budget optimization, both of which are critical for maximizing return on investment [38, 95, 96].

All the above KPIs are snapshots of a point in time. Advertisers, however, need to be able to assess the long-term effects of their efforts. Metrics which go beyond immediate ad or campaign performance or financial metrics to provide brand-related indications of the impact of PA are the following:

(g) Brand Recall and Brand Recognition evaluate the lasting impressions an advertisement makes. Because PA affords optimized frequency and targeting it is more likely to establish higher levels of recall and recognition, more durable brand connections, and better streamlined brand associations than traditional advertising [97–99].
(h) Neuromeric measure emotional engagement by employing sophisticated tools, such as eye-tracking, Facial Expression Analysis (FEA), and attention and emotional response analysis using voice, electrocardiogram, electrodermal activity, respiration, and other biometrics. Such technologies, which are used in laboratory settings but can also be used remotely, using web-based software and the users' laptop or mobile device webcam, provide detailed measurements of the users' cognitive and emotional reactions to advertisements. Such metrics can help refine the creative elements of a campaign in a way that amplifies emotional but also cognitive resonance [36, 100, 101]. As with all advertising efforts, financial metrics establish a direct relationship between advertising expenditure and revenue outcomes. The following metrics serve as essential tools for measuring profitability and supporting resource allocation decision-making.
(i) Return on Ad Spend (ROAS) quantifies the revenue generated per dollar spent on advertising. This makes it a key performance indicator (KPI) of financial efficiency. PA platforms use machine learning algorithms to optimize ROAS in real-time by reallocating budget resources to high-performing audience segments [38, 102]
(j) Cost Per Mille (CPM) calculates the cost per thousand impressions and is commonly used to assess financial efficiency in campaigns with broad awareness objectives.

While primarily associated with visibility-focused initiatives, CPM also holds relevance for performance-driven campaigns where reach supports conversions. By comparing CPM across platforms, advertisers can evaluate cost efficiency relative to audience reach [6, 7].

Finally, there is a need to track PA effectiveness throughout the consumer journey. The technologies employed in the PA ecosystem afford the data, computing power and advanced methodologies necessary to construct, track, and evaluate KPIs such as:

(k) Attribution Modelling which can be Multi-Touch Attribution (MTA), which allocates credit across various touchpoints in the customer journey to provide a comprehensive understanding of how different interactions contribute to conversions, or Last-Click Attribution (LCA) which assigns credit to the final interaction. Although easier to interpret and more cost-effective, LCA can result in underestimating the role of earlier touchpoints in driving brand awareness or behavioral intent. Selecting the most appropriate attribution model is critical for aligning measurement techniques with campaign objectives [103–105].
(l) Incrementality Testing evaluates the impact of PA by comparing outcomes between users exposed to ads and a control group not exposed. Incrementality testing isolates the advertising effect from organic behaviors, thus providing a clear assessment of the true value of the PA campaign [106, 107]. For advertisers with limited budgets, this approach ensures resources are allocated to strategies that demonstrably drive incremental results.

It is obvious that these KPIs are but tools which can be used to balance PA reach, visibility, cost, and efficacy. Some are simple to construct and interpret, others are more sophisticated and robust measures of the overall value of the advertising effort of the organization. To get a clear picture, a combination of these approaches needs to be employed as part of the strategic decision-making process for long-term marketing success.

6 To PA or Not to PA? That is the Question for SMEs

It is obvious from the above presentation of PA that employing it is not a simple, straightforward, intuitive process which is accessible to anyone with a basic business and marketing skillset. It is not at all like traditional ad buying, where the advertiser and the agency feel in control of a process they have used and seen being used numerous times. The speed and level of automation of PA can leave the advertiser feeling at a loss and wondering if they have made the right choices. Like with everything else in business decision-making, to choose between alternative paths you need to know how the different options work, their pros and cons, their benefits and potential pitfalls. Not many marketers and small business owners, however, can safely say they do know how PA works, how their data is being used and by whom, who gets paid and how, and how reliable the metrics they get from the platforms are. If you look at PA from the perspective of an SME it becomes obvious that, if anything, it is the benefits of PA which actually pose the most significant challenges of using it. We here present the issues German SME owners and marketing decision-makers identified as the key benefits and major hurdles in PA adoption and full exploitation alongside the relevant literature.

6.1 Benefits of PA for SMEs

Probably the most pressing problem for SMEs is to elaborate an appealing marketing message which cuts through the media clutter and gets across to those who need to receive it, at a time that matters to them. This is intensified by the fact that they have CRM systems which are either ad-hoc, or, usually, underdeveloped and lacking technological sophistication, limited advertising budgets, little or no in-house marketing expertise, and a size that precludes them becoming the clients of the big multinational advertising and digital marketing agencies.

The unparalleled targeting precision PA offers is the key reason for using it. Our informants, who were business owners, CMOs, and online marketing managers from both agencies and retail companies, highlighted the following advantages of Programmatic Advertising (PA) with respect to targeting:

a) Real time optimization of advertisements considerably improves targeting precision compared to traditional advertising or manually implementing digital advertising.
b) PA enables companies to specifically reach customers who are genuinely interested in their products thus reducing ad waste and improving ad efficiency.
c) PA increases reach without sacrificing targeting precision and efficiency.

Although not cheap, PA is perceived as cost effective provided sufficient financial resources are available. The key financial benefits they attributed to PA were the following:

d) Real-time control of budgets effectively reduces waste and eliminates overextending the pre-determined budget.
e) PA optimizes expenditures without sacrificing reach.
f) Automated adjustments save valuable time, particularly for companies lacking the capacity to manage campaigns manually.

An important issue that was raised by the executives and account managers of SMEs who are digital marketing and advertising agencies was the growth potential of PA, especially in view of emerging market trends and technological advancements. Key benefits they identified were the following:

g) AI-supported automation solutions such as ad creative tools, or applications which can perform automated A/B and playout testing, can be integrated into existing tools and techniques to create more appealing and highly personalized ads.
h) The increasing accessibility and transparency of PA and AI technologies could unlock significant market development opportunities through automated generation of ad creatives which could further streamline and optimize marketing processes.
i) Advances in artificial intelligence (AI), machine learning, and big data are fundamentally transforming the PA landscape, enabling greater targeting precision, more sophisticated audience segmentation, and real-time optimization, but, most importantly, they minimize the need for extensive human oversight, making PA both scalable and accessible to businesses of various sizes.

Finally, SME owners and managers as well as agency executives agreed that:

j) PA improves both the efficiency and the effectiveness of advertising efforts in the digital domain and complements rather than cannibalizes marketing communications, sales, and customer service in the physical domain creating a seamless phygital marketing environment.

6.2 Challenges and Barriers to PA Implementation by SMEs

Not all is rosy in the PA implementation arena for SMEs, however. We have already pointed out financial [11–13, 37], technological [1, 108], skills [38, 78, 109], and regulatory [67, 68] barriers through the literature review presented above. Germann SME owners and CMOs, and online marketing managers saw, more or less, the same issues standing in the way of their fully implementing the cutting-edge technologies available to them. As many SMEs perceive programmatic advertising as a 'black box' (see point 6.2.e), transparent and simple training measures (see Table 3, column 4) should be implemented.

The key financial challenges they identified were:

a) High setup costs and considerable internal effort required.
b) After running a few campaigns, it often becomes evident that achieving substantial results with limited budgets is challenging.
c) SMEs with limited budgets struggle to achieve adequate reach.

Many of the SME owners we talked to when writing this chapter described the systems as a "black box," the functioning of which they could not fully comprehend. The technology and skills gaps were seen as a single issue which, however, raised multifaceted concerns such as the following:

d) Business owners, CMOs, and online marketing managers from agencies and retail companies positioned the lack of technical expertise required to manage the PA system internally as the primary, if not insurmountable barrier and the cause for not reaping the benefits of digital technologies to the maximum. Indeed, programmatic advertising for SMEs.
e) Lack of technical expertise is associated with a perception of limited transparency of the PA ecosystem and its operation. SME managers who felt they lacked the knowledge to implement a PA system also highlighted transparency concerns as a key obstacle.
f) A perception of the firm as of low technological sophistication leads to limited trust in PA technologies, service and outsourced infrastructure providers and digital marketing and advertising agencies, particularly at the management level.
g) The complexity of the technical infrastructure required for PA was perceived as a significant challenge for firms who felt that operating such platforms would be nearly impossible without specialized IT personnel which they were worried would be hard to find or afford.
h) SMEs often have insufficient overall technical maturity.
i) SMEs who tried to their hand in PA soon found that they had overestimated their internal capabilities which proved inadequate for successfully implementing programmatic advertising in the long run.

j) There is often a lack of understanding of the technical and operational requirements of PA.

Finally, the evolving regulatory environment is seen as presenting SMEs which have some experience with PA and small agencies with a number of challenges such as:

k) GDPR complicates the effective use of PA.
l) There is a widespread fear of penalties in cases of non-compliance and many companies are concerned about potential legal consequences or find it too complex to use programmatic advertising in full compliance with data protection regulations.
m) The GDPR regulations regarding the phasing-out of third-party cookies present specific challenges for PA implementation and limit its efficiency in delivering personalized advertisements.

As SMEs face high setup costs when implementing PA, making it difficult to achieve sufficient reach (Challenge 6.2.b). This financial constraint for instance is addressed in Table 3 (Recommendation 2), which suggests leveraging funding options and strategic budget allocation to optimize resources. Similarly, compliance with GDPR (Challenge 6.2.k) is a barrier that Table 3 (Recommendation 1) addresses by recommending a cost-effective data protection strategy.

Dilemmas are oversimplifications of the lived reality of businesses, especially of the pressing conditions faced by SMEs who have to compete with MNCs and large corporations in the digital domain and even in their immediate physical neighborhood. Moreover, technological progress is non-reversable, so, sooner or later, even the smallest companies will have to find their way in the complex technological domain of business automation. So, the real question is not to PA or not to PA but how to reap the significant benefits enjoyed by the larger competitors. To this end, we propose the following Four-Phase Framework for PA Implementation.

7 Four-Phase Framework for PA Implementation

PA implementation can be developed in four phases which are distinct but interrelated and work in a virtuous cycle manner of analysis-planning-implementation-evaluationanalysis, etc. Each phase has a specific objective which is met by completing a set of actions. We also offer some actionable recommendations for steps to take and resources to utilize.

7.1 Phase 1: Analysis

Objective: To assess organizational readiness and prerequisites. The related actions involve resource assessment, market analysis, and regulatory compliance assessment.

The **resource assessment** step involves the evaluation of technical infrastructure and. Staff expertise using tools such as resource checklists or diagnostic software and setting a budget. The outcome of the resource assessment process is a comprehensive identification of strengths, areas for improvement, and functions which should be outsourced. **Market analysis** involves competitor and target audience analysis to define target audiences and preferred advertising channels. **Regulatory compliance assessment** is needed to ensure that all data collection and management processes and infrastructure components adhere to data protection regulation such as GDPR. It is likely that at this phase, SMEs will need to seek external expertise if internal resources are insufficient. In this phase, we recommend the following practical steps and resources:

- Use free analytics tools
  - Google Analytics™, Meta Business Suite™ for audience insights.
  - SimilarWeb™ (free version) for competitor research.
- Check Regulatory compliance with minimal cost
  - Assess GDPR compliance using checklists from local business chambers.
  - Consult IHK or GDPR specialists for basic data compliance checks
  - Test free Consent Management Platforms (CMPs) such as OneTrust™
- Start with a small feasibility test:
  - Allocate €500–€1,000 to a low-risk test campaign.
  - Use retargeting ads to target existing customers first.

7.2 Phase 2: Planning

Objective: To develop a strategy and allocate resources. This involves three sets of interrelated actions which are very similar to all advertising campaign design and planning. A measurable use case should be selected as a pilot project. For example, a local retailer might choose to promote seasonal discounts via a PA strictly localized regional campaign. **Pilot project design** should be kept simple. **Budgeting and scheduling** involve establishing a clear timeline and allocating resources. It is recommended that between 10 and 20% of the annual marketing budget is allocated to PA testing. For the pilot PA project small budgets can be allocated to Google Display Ads™ or Meta Ads™. It is recommended that only one or two key segments (e.g., existing customers and lookalike audiences are targeted. **Technology and partner selection** require the evaluation of alternative PA platforms (e.g., Google Ads™, or Trade Desk™) or external full-service providers based on expertise, user-friendliness, cost, and suitability for the specific test campaign. In this phase, we recommend the following practical steps and resources:

- Leverage automated bidding strategies for better budget control.
- Outsource instead of building an in-house PA team

  Hire freelancers on Upwork or Fiverr for initial PA campaign setup.
  Work with a small digital agency on a single ad goal before committing long-term.

- Set realistic performance benchmarks
  - Aim for at least a 3% CTR and 1% conversion rate.
  - Compare PA results against existing Google or social ad performance.

7.3 Phase 3: Implementation

Objective: To execute the pilot project and gather insights. This phase comprises the following tasks:

1. Employee training through online modules or workshops
2. Pilot launch which involves the execution of the campaign as planned and the monitoring of its chosen performance metrics (e.g. click-through or conversion rates
3. Real-time adjustments to optimize the campaign based on collected data such as modifying audience targeting or content strategies.

For companies without an in-house digital marketing team the same steps are carried out with the assistance of a specialized, preferably boutique agency or freelancer. In this phase, we recommend the following practical steps and resources:

- Use cost-efficient automation tools:
  - Google Ads Smart Bidding™ for automated bid management.
  - Canva™ or Creatopy™ for affordable ad creatives.
- Use simple tracking methods instead of expensive DMPs:
  - Set up Google Tag Manager™ (free and easy for SMEs).
  - Leverage first-party data by growing a newsletter database.
- Opt for budget-friendly DSP alternatives. Instead of Google DV360, SMEs can use:
  - The Trade Desk Express™ (SME-friendly)
  - StackAdapt™ for small-budget display ads

7.4 Phase 4: Evaluation

Objective: To assess the pilot PA project success and decide on scaling. To harness the data required for informed and rational decision making on PA adoption, SMEs need to carry out **performance analysis**, which involves evaluating KPIs, including ROI and cost per lead and conversion using reporting dashboards, and maintain full **documentation** of successes, challenges, and lessons learned for future campaigns. On the basis of these, they are called to make **scaling decisions** aiming to gradually expand PA activities based on pilot outcomes and the integrate of the test PA findings into the long-term marketing strategies. To meet the evaluation phase objectives we recommend:

- Using simple KPI dashboards (e.g. Google Looker Studio™) to visualize performance data.
- Adopt simple, meaningful KPIs instead of complex analytics
  - Focus on CTR (Click-Through Rate) and Conversions.
  - Ensure cost per acquisition (CPA) is no more than 20–30% of product value.

- Only scale if ROI is positive.
  - If CPA is lower than the expected customer lifetime value (CLV) double the budget.
  - If CPA is too high, pause and optimize before reinvesting.
- Train internal teams using cost-free PA learning tools
  - Use Google Skillshop™ & Meta Blueprint™.
- Provide basic PA training for key employees before expanding efforts. Moreover, based on the analysis of the literature and the observations of the key barriers to PA adoption SME owners and managers shared with us we offer the following

Targeted recommendations for facilitating PA adoption and present their rationale in Table 3.

Table 3. Targeted recommendations for facilitating PA adoption by SMEs

Challenge	Recommendation	Rationale	Reference to 6.2
Technical infrastructure and data protection	Establish a robust IT infrastructure with a data protection management system to ensure GDPR compliance. Appoint a Data Protection Officer (DPO) or collaborate with specialized providers for data governance	Addressing data protection challenges builds customer trust and reduces compliance risks	6.2.k, 6.2.l, 6.2.m – GDPR requirements complicate data use and add compliance risks
Budget allocation and resource planning	Develop long-term budget strategies, leveraging grants or partnerships with advertising platforms to optimize resource use	Financial constraints hinder adoption. Strategic investments in automation and efficient tools lower operational costs and enhance campaign performance	6.2.a, 6.2.b, 6.2.c – High setup costs, difficulty in achieving sufficient reach with limited budgets
Workforce development and training	Provide specialized training to staff on PA technologies and outsource complex tasks to expert marketing agencies as needed	Addressing skill gaps accelerates adoption and improves outcomes, mitigating the challenge of limited internal expertise	6.2.d, 6.2.e, 6.2.f, 6.2.g, 6.2.h, 6.2.i, 6.2.j – Lack of technical expertise, per opacity of PA, limited trust in service providers
Marketing strategy and positioning	Develop an integrated marketing roadmap that combines datadriven advertising with clear branding strategies	Precise targeting through data-driven techniques, coupled with strong branding, enhances market competi-tiveness	6.2.d, 6.2.g, 6.2.j – SMEs lack technical maturity and a clear understanding of PA requirements
Technology adoption and innovation management	Conduct pilot campaigns to test PA systems on a smaller scale and collaborate with technology providers for efficient implementation	Gradual adoption reduces risk and fosters suc integration of innovative solutions	6.2.h, 6.2.i, 6.2.j – SMEs struggle with overestimating internal capabilities and technical challenges

8 Conclusion

Advances in artificial intelligence (AI), machine learning, and big data are fundamentally transforming the programmatic advertising (PA) landscape, enabling greater targeting precision, more sophisticated audience segmentation, and real-time optimization. Most importantly, these technologies reduce the need for extensive human oversight, making PA both scalable and accessible to businesses of various sizes [3, 110]. As AI-driven automation continues to evolve, PA is becoming increasingly efficient and cost-effective, offering advertisers the ability to optimize campaigns dynamically and personalize engagement at an unprecedented scale.

Beyond its current capabilities, PA is expanding into emerging channels such as programmatic television, audio advertisements, and immersive media formats. These developments create new opportunities for businesses to engage consumers in more interactive and personalized ways. For instance, programmatic TV allows advertisers to deliver targeted messages to specific households, while digital-out-of-home (DOOH) advertising leverages real-time data to enhance audience engagement. Furthermore, augmented reality (AR) and virtual reality (VR) technologies introduce immersive advertising experiences that captivate consumers and foster brand loyalty [8, 111]. These advancements underscore PA's ability to meet growing consumer expectations for relevant and engaging ad content.

Despite its transformative potential, the adoption of PA - particularly among small and medium-sized enterprises (SMEs) - remains hindered by challenges such as high setup costs, technical complexities, and regulatory compliance requirements. However, these barriers are surmountable with strategic investments in infrastructure, workforce development, and ethical data management. Implementing structured adoption frameworks, such as the Four-Phase Framework proposed in this chapter, can help businesses effectively navigate the complexities of PA while maximizing its benefits. As the regulatory landscape continues to evolve, particularly in the areas of data privacy and consumer protection, businesses must prioritize compliance while maintaining agility in adopting new marketing technologies. The shift toward first-party data strategies and ethical data practices will be crucial in fostering consumer trust and ensuring long-term sustainability.

Future research should explore tailored strategies for businesses, focusing on first-party data utilization, ethical data management, and innovative targeting methods. Longitudinal studies on PA's impact on competitiveness and sustainability would offer valuable insights for academics and practitioners alike. By addressing these challenges and capitalizing on emerging opportunities, businesses can position PA as a pivotal driver of digital transformation, ensuring resilience and success in a datadriven marketplace.

Appendix: Key Terms of Programmatic Advertising

An overview of the key PA terms with their definition and notes on each one's purpose, importance and key metrics is presented below to further support SMEs in effective decision-making towards implementing cost-effective PA strategies.

Term	Definition	Purpose, Importance and Key Metrics
(Ad) Impression	Counting an ad display to a user, a basic metric for measuring ad views	Indicates the ad's relevance and engagement with the target audience, as a high CTR suggests strong alignment between the ad content and user interests. CTR is essential for assessing the appeal of ad creative and messaging
(Ad) Inventory	Ad spaces available on publishers' websites for purchase via programmatic channels	Measures the effectiveness of the ad in driving meaningful user actions, such as purchases or subscriptions, after clicking. A high conversion rate indicates that the ad resonates strongly and drives tangible business outcomes
Ad Exchange	Marketplace for exchanging ad placements between buyers and sellers	Supports budget management and efficiency by tracking the cost per indi click, enabling advertisers to monitor ad spend and optimize costeffectiveness in engaging potential customers
Advertisers	Marketers who purchase ad placements to reach target audiences	A valuable metric to evaluate the financial efficiency of ad exposure by measuring the cost per thousand impressions, indicating how much is spent to reach a broader audience with targeted messaging
Audience Reach	Number of unique viewers who saw the ad, measuring reach and targeting effectiveness	Directly assesses the campaign's financial efficiency by evaluating the revenue generated for each dollar spent on advertising, guiding budget allocation based on return on investment
Click-Through Rate (CTR)	Percentage of ad clicks in relation to its impressions	Essential for determining the quality of impressions in programmatic platforms by measuring the percentage of ads that were actually viewable to users, thus ensuring ads reach users effectively

(*continued*)

(*continued*)

Term	Definition	Purpose, Importance and Key Metrics
Conversion Rate (CVR)	Proportion of users who complete a desired action (e.g., purchase) after clicking on the ad	Tracks deeper levels of user interaction, such as shares, video plays, or social engagements, providing insights into user interest and engagement beyond mere clicks, valuable in awareness oriented campaigns
Cookies	Small files stored on a user's computer that record information about previous website visits	Measures the scope of unique users exposed to an ad, serving as an indicator of reach and ensuring that ads are successfully delivered to the intended demographic or psychographic audience
Cost Per Acquisition (CPA)	Cost of acquiring a customer or a specific action, essential for budget allocation	Ensures the ad maintains visibility without causing user fatigue by measuring the frequency of ad exposure to the same user, supporting a balanced approach between brand visibility and user experience
Cost Per Click (CPC)	Cost per ad click, used for budget management and cost-efficiency analysis	Helps in understanding the acquisition costs for each customer or targeted action, allowing advertisers to manage budgets more effectively and focus on strategies that yield cost-efficient customer acquisition
Cost Per Mille (CPM)	Cost per thousand impressions, measuring advertiser cost for ad reach and visibility	Increases the efficiency of digital campaigns by automating the process of media buying, enabling precise targeting, and optimizing ad placements in real-time, which enhances reach and budget effectiveness

(*continued*)

(*continued*)

Term	Definition	Purpose, Importance and Key Metrics
Data Management Platform (DMP)	Platforms supporting DSPs with data acquisition and targeting enhancement	Facilitates precise ad placement to target audiences by enabling advertisers to bid for impressions in real-time auctions, increasing the likelihood of engaging relevant users with tailored ads
Demand Side Platform (DSP)	Platforms assisting advertisers in selecting and purchasing impressions	Provides a fundamental metric to count ad exposures to users, serving as the baseline for assessing overall campaign reach and tracking the visibility of the advertisement
Display Advertising	Online advertising, mainly in banner form, delivering marketing messages to website visitors	Defines the available ad spaces on publishers' platforms, representing the inventory that can be programmatically acquired by advertisers for efficient ad placement
Engagement Rate	Metrics that measure user interactions (e.g., shares, video views), crucial for awareness campaigns	Aims to communicate targeted messages to web visitors by displaying banner ads, which can drive awareness and brand recall among users browsing digital spaces
Frequency	Number of exposures of the same ad to the same user within a given period, preventing oversaturation	Blends into the visual format of the hosting platform, offering a seamless and less intrusive experience for users and enhancing the likelihood of engagement by matching platform aesthetics
Publishers	Operators of websites providing ad spaces for advertisers	Assists advertisers in streamlining the ad placement process by selecting and purchasing the best-suited impressions, which optimizes targeting precision and budget efficiency

(*continued*)

(*continued*)

Term	Definition	Purpose, Importance and Key Metrics
Real-Time Bidding (RTB)	Process to deliver ads to targeted audiences through real-time auctions	Helps publishers effectively manage their ad spaces, maximizing ad revenue while ensuring suitable ads are displayed to users in alignment with publisher and audience interests
Retargeting	Repeated advertisement exposure for users who previously interacted with similar content	Facilitates an automated and streamlined transaction process for ad placements, allowing real-time bidding for efficient exchange between buyers and sellers in the advertising marketplace
Return on Ad Spend (ROAS)	Revenue generated relative to advertising spend, assessing financial efficiency	Enables DSPs to improve targeting capabilities by gathering and analyzing data from multiple sources, which assists advertisers in reaching audiences based on specific criteria
Supply Side Platform (SSP)	Platforms helping publishers manage and sell their ad spaces	Personalizes ad messages to appeal to users based on collected personal information, like demographics and purchase history, which enhances ad relevance and the likelihood of user engagement
Targeting	Targeted advertisements based on personal data, such as name, purchase history, demographics, and interests	Boosts conversion likelihood by re-engaging users who previously showed interest, using past browsing behaviors to display relevant ads that prompt continued interaction with the brand
Viewability Rate	Percentage of ads actually viewable to users, ensuring effective impressions	Allows collection of user behavior data to improve targeting and retargeting accuracy, facilitating better-tailored advertisements that align with user preferences and prior online activities

Source: [11, 13, 35, 74, 84, 86, 88, 94, 96, 97, 102, 104, 112–121]

References

1. Li, H.: My experience in teaching programmatic advertising. J. Advert. Educ. **23**(2), 100–107 (2019)
2. Interactive Advertising Bureau: Attitudes to Programmatic Advertising Report 2024. Programmatic-Advertising-Report-Nov-2024.pdf (2024)
3. McStay, A.: 3.1 micro-moments, liquidity, intimacy and automation: developments in programmatic ad-tech. In: Siegert, G., Rimscha, M.B., Grubenmann, S. (eds.) Commercial Communication in the Digital Age, pp. 143–160. De Gruyter. Berlin (2017)
4. Hasri, M., Silva, N., Susanto, Y., Winarno, M.P.: The attitude toward the programmatic advertising on click-through intention. In: Proceedings of the 4th International Conference of Economics, Business, and Entrepreneurship, ICEBE 2021 (2022)
5. Bundesverband Digitale Wirtschaft e.V.: Programmatic Advertising Kompass 2015/2016. BVDW: Duesseldorf, p. 118 (2015)
6. Bundesverband Digitale Wirtschaft e.V.: Programmatic Advertising Kompass 2016/2017 [Programmatic Advertising Compass 2016/2017]. 2nd ed. BVDW. Duesseldorf: BVDW (2016)
7. Bundesverband Digitale Wirtschaft e.V.: Programmatic advertising kompass 2017/2018 [Programmatic Advertising Compass 2017/2018]. 3rd ed. BVDW. Duesseldorf: BVDW (2017)
8. Meirezaldi, O.: Programmatic advertising: evolution, efficacy, and ethical implications in the digital age. J. Publicuho **6**(3), 1069–1079 (2023)
9. Jensen, H., Sund, K.J.: The implications of programmatic advertising on the business model of TV broadcasters. In: Khajeheian, D., Friedrichsen, M., Mödinger, W. (eds.) Competitiveness in Emerging Markets: Market Dynamics in the Age of Disruptive Technologies, pp. 35–53. Springer, Heidelberg (2018)
10. Ungureanu, A., Popescu, D.-L.: Online advertising history, evolution, and challenges. Revista Economica **74**(3), 121–130 (2022)
11. Busch, O.: Realtime Advertising: Digitales Marketing in Echtzeit: Strategien, Konzepte und Perspektiven [Realtime Advertising: Digital Marketing in Real Time: Strategies, Concepts and Perspectives]. Springer Gabler (2014)
12. Busch, O.: The programmatic advertising principle in programmatic advertising: the successful transformation to automated. In: Busch, O. (ed.) Data-Driven Marketing in Real-Time, pp. 3–15. Springer, Cham (2015)
13. Busch, O.: Programmatic Advertising: The successful transformation to Automated, Data-Driven marketing in Real-Time. Management for Professionals. Springer Gabler (2016)
14. Cannella, J.: Artificial Intelligence In Marketing [Honors thesis, Arizona State University] (2018)
15. Li, J., Yuan, Y., Zhao, X., Wang, F.-Y.: Research on information structure of programmatic advertising markets. IFAC-PapersOnLine **50**(1), 13587–13592 (2017)
16. Funk, B., Nabout, N.A.: Cross-channel real-time response analysis in programmatic advertising: the successful transformation to automated. In: Busch, O. (ed.) Data-Driven Marketing in Real-Time, pp. 141–151. Springer, Cham (2016)
17. Gertz, O., McGlashan, D.: Consumer-centric programmatic advertising in programmatic advertising: the successful transformation to automated. In: Busch, O. (ed.) Data-Driven Marketing in Real-Time, pp. 55–73. Springer, Cham (2016)
18. Malthouse, E.C., Maslowska, E., Franks, J.U.: Understanding programmatic TV advertising. Int. J. Advert. **37**(5), 769–784 (2018)
19. Setkute, J., Dibb, S.: From theory to practice: practical implications as a translational bridge between research relevance and impact. Ind. Mark. Manag. **125**, 131–149 (2025)

20. Benady, D.: Programmatic: packed with potential and peril. In: Catalyst, Charted Institute of Marketing, pp. 13–19 (2015)
21. Samuel, A., White, R.T.G., Thomas, R., Jones, P.: Programmatic advertising: An exegesis of consumer concerns. Comput. Hum. Behav. **116**, 106657 (2021)
22. Rudnicka, J.: Mittelstand als Rückgrat der Wirtschaft | Statista [SMEs as the Backbone of the Economy|Statista]. https://de.statista.com/themen/3996/mittelstand-in-deutschland/#topicOverview. Accessed 20 March 2024
23. Bang, H., Kim, J., Choi, D.: Exploring the effects of ad-task relevance and ad salience on ad avoidance: the moderating role of internet use motivation. Comput. Hum. Behav. **89**, 70–78 (2018)
24. Palos-Sanchez, P., Saura, J.R., Martin-Velicia, F.: A study of the effects of programmatic advertising on users' concerns about privacy overtime. J. Bus. Res. **96**, 61–72 (2019)
25. Priebe, A.: Rückblick auf 10 Jahre Programmatic in Deutschland [A look back at 10 years of programmatic in Germany]. https://www.adzine.de/2023/12/rueckblick-auf-10-jahre-programmatic-in-deutschland. Accessed 12 Dec 2023
26. Aslam, B., Karjaluoto, H.: Digital advertising around paid spaces, E-advertising industry's revenue engine: a review and research agenda. Telemat. Inform. **34**(8), 1650–1662 (2017)
27. Bundesverband Digitale Wirtschaft e.V.: Umsätze weiter auf Wachstumskurs [Sales continue to grow]. https://www.bvdw.org/news-und-publikationen/umsaetze-weiter-auf-wachstumskurs/. Accessed 06 Mar 2024
28. Keller, D., Stavenhagen, L., Eggert, C.: Zur Wirtschaftlichen Lage Der Deutschen Zeitungen 2020 [On the Economic Situation of German Newspapers in 2020] (2020). https://www.bdzv.de/fileadmin/content/7_Alle_Themen/Marktdaten/Bericht_Umsatz erhebung_Herr_Keller_V2.pdf
29. Adzine: Microsoft nimmt Abschied von der Marke Xandr [Microsoft bids farewell to the Xandr brand]. https://www.adzine.de/2023/06/microsoft-nimmt-abschied-von-der-marke-xandr/. Accessed 20 June 2023
30. The Nielsen Company: EXelate – Data as a service (2017). https://microsites.nielsen.com/daas-partners/partner/exelate/
31. Hammouri, I., Ameri, J., Maltijian, D., AlKarablieh, M., Ismail, H.A.B.: Evolution of design in Advertising (From traditional to Digital) embraced by industry professionals. J. Tour. Hosp. Sports **42**, 51–66 (2019)
32. Su, N., Luo, L.: Historical track of the development of media technologies and the evolution of advertising forms. In: Proceedings of the 2015 4th International Conference on Sensors, Measurement and Intelligent Materials, pp. 474–479 (2016)
33. Kollmann, T.: Handbuch Digitale Wirtschaft [Handbook Digital Economy]. 1st ed. Handbuch Digitale Wirtschaft. Wiesbaden, Germany: Springer Gabler. 1338 (2020)
34. Kollmann, T.: Einführung in das Online-Marketing [Introduction to online marketing]. In: Kollmann, T. (ed.) Handbuch Digitale Wirtschaft, pp. 1–27. Springer Gabler, Wiesbaden (2018)
35. Yuan, Y., Wang, F., Li, J., Qin, R.: A survey on real time bidding advertising. In: Proceedings of 2014 IEEE International Conference on Service Operations and Logistics, and Informatics, pp. 418–423 (2014)
36. Zhang, S., Wakefield, R., Huang, J., Li, X.: Exploring determinants of consumers' attitudes toward real-time bidding (RTB) advertising. Inf. Technol. People **34**(2), 496–525 (2020)
37. Seitz, J., Zorn, S.: Perspectives of programmatic advertising in programmatic advertising: the successful transformation to automated. In: Busch, O. (ed.) Data-Driven Marketing in Real-Time, pp. 37–51. Springer, Heidelberg (2015)
38. Bishop, T.: As programmatic advertising becomes the new normal, how can advertisers create greater consumer engagement and publishers ensure greater return? J. Dig. Social Media Mark. **5**(1), 6 (2017)

39. Martínez-Martínez, I.J., Aguado, J.-M., Boeykens, Y.: Ethical implications of digital advertising automation: the case of programmatic advertising in Spain. El Profesional de la Información **26**(2), 201–210 (2017)
40. O'Hara, C.: Data Triangulation: How Second-Party Data Will Eat The Digital World.https://www.adexchanger.com/data-driven-thinking/data-triangulation-how-second-party-data-will-eat-the-digital-world/. Accessed 25 Jan 2016
41. Guitart, I.A., Hervet, G., Gelper, S.: Competitive advertising strategies for programmatic television. J. Acad. Mark. Sci. **48**(4), 753–775 (2020)
42. Panigyrakis, G.G., Zarkada, A.K.: A philosophical investigation of the transition from Integrated Marketing Communications to metamodern meaning cocreation. J. Glob. Scholars Market. Sci. **24**(3), 262–278 (2014)
43. Zarkada, A.K.: What makes a B2B Facebook page engaging? In: 2020 Global Marketing Conference at Seoul Bridging Asia and the World: New Marketing and Management in Digitally Connected World Virtual Conference: Global Alliance of Marketing & Management Associations (2020)
44. Zarkada, A.K.: Marketing solutions to the challenges of a VUCA environment: what have we accomplished and what more can we do? In: Marketing Solutions to the Challenges of a VUCA Environment. Springer, Cham (2024)
45. Thomas, G.: Data usage in talent management – challenges for SMEs in the field of skilled crafts. SocioEcon. Chall. **4**(1), 75–81 (2020)
46. Linden, T., Khandelwal, R., Harkous, H., Fawaz, K.: The privacy policy landscape after the GDPR. Cryptogr. Secur. (3) (2019)
47. J.O.M. Group: Ende der Third Party Cookies: Aktuelle Entwicklungen & Alternativen für ein datenschutzfreundliches Marketing [The end of third-party cookies: current developments & alternatives for privacy-friendly marketing]. https://www.jom-group.com/ende-der-third-party-cookies-aktuelle-entwicklungen-alternativen-fuer-ein-datenschutzfreundliches-marketing/. Accessed 07 Mat 2023
48. Maurer, B.: Ende der third party cookies & die auswirkungen auf social media [End of third party cookies & the impact on social media] (2025). https://www.506.ai/das-ende-der-third-party-cookies/
49. Ponemon Institute: Data Risk in the Third-Party Ecosystem (2016). https://www.ponemon.org/local/upload/file/Data%20Risk%20in%20the%20Third%20Party%20Ecosystem_BuckleySandler%20LLP%20and%20Treliant%20Risk%20Advisors%20LLC%20Ponemon%20Research%202016%20-%20FINAL2.pdf
50. Witthaus, A.: Keine Third-Party-Cookies mehr – so lecker sind die neuen Targeting-Trends [No more third-party cookies - this is how delicious the new targeting trends are]. https://planning.doccheck.com/articles/keine-third-party-cookies-mehr-so-lecker-sind-die-neuen-targeting-trends
51. Meyer, M.: Das Ende der 3rd Party Cookies [The end of 3rd party cookies]. https://www.mso-digital.de/blog/das-ende-der-3rd-party-cookies/. Accessed 18 Mar 2024
52. Laufer, W.S.: Integrity, diligence and the limits of good corporate citizenship. Am. Bus. Law J. **34**, 157–181 (1996)
53. Carroll, A.B.: The four faces of corporate citizenship. Bus. Soc. Rev. (100/101), 1–7 (1998)
54. Maignan, I.S., Ferrell, O.C.: Corporate citizenship: cultural antecedents and business benefits. J. Acad. Mark. Sci. **27**(4 Fall), 455–469 (1999)
55. Mackey, S.: Virtue ethics, CSR and "corporate citizenship." J. Commun. Manag. **18**(2), 131–145 (2014)
56. Lutz, C., Hoffmann, C.P., Ranzini, G.: Data capitalism and the user: an exploration of privacy cynicism in Germany. New Media Soc. **22**(7), 1168–1187 (2020)
57. Härting, N.: Study on the Impact of the Proposed ePrivacy Regulation (2017). https://www.informationpolicycentre.com/uploads/5/7/1/0/57104281/epr_-_gutachten-final-4.0_3_.pdf

58. GDPR EU: GDPR: General Data Protection Regulation (2025). https://www.gdpreu.org/
59. Krämer, A., Mauer, R.: Datenschutz für Entscheider in Marketing und Vertrieb [Data protection for decision-makers in marketing and sales]. Springer Gabler, Wiesbaden (2023)
60. Trenkle, J.: Digital Transformation in Small and Medium-Sized Enterprises: Strategy, Management Control, and Network Involvement, 1st ed. Wirtschaft und Recht für Mittelstand und Handwerk. Vol. 3. Baden-Baden: Nomos (2020)
61. Martin, K.D., Murphy, P.E.: The role of data privacy in marketing. J. Acad. Mark. Sci. **45**(2), 135–155 (2016)
62. Interactive Advertising Bureau: GDPR impact on Programmatic Trading blog series: IAB Europe's new GDPR Transparency and Consent Framework – a unique opportunity for publishers. https://iabeurope.eu/gdpr-impact-on-programmatic-trading-blog-series-iab-europes-new-gdpr-transparency-and-consent-framework-a-unique-opportunity-for-publishers/. Accessed 12 Apr 2018
63. McCarthy, J.: IAB Europe falls foul to GDPR, advertisers may have to destroy data. https://www.thedrum.com/news/2022/02/02/iab-europe-falls-foul-gdpr-which-may-result-many-advertisers-having-destroy-data. Accessed 02 Feb 2022
64. CMIT Solutions: Protect Your Business with GDPR Compliance (2025). https://cmitsolutions.com/charleston-sc-1165/blog/gdpr-compliance-for-businesses-data-protection-in-the-digital-age/
65. Geradin, D., Katsifis, D., Karanikioti, T.: GDPR myopia: how a well-intended regulation ended up favoring google in ad tech. TILEC Discussion Paper (2020)
66. Batista, M., Fernandes, A.L., Ribeiro, L.P., Alturas, B., Costa, C.P.: Tensions between privacy and targeted advertising. Sci. J. Appl. Social Clin. Sci. **3**(14), 2–7 (2023)
67. Cooper, D.A., Yalcin, T., Nistor, C., Macrini, M., Pehlivan, E.: Privacy considerations for online advertising: a stakeholder's perspective to programmatic advertising. J. Consum. Mark. **40**(2), 235–247 (2023)
68. Hanáková, L.: Current issues of personal data privacy in marketing. Acta academica karviniensia **20**(2), 5–20 (2020)
69. Buske, F.: Ohne Daten kein Programmatic Advertising setzt die DSGVO automatisierter Werbung jetzt ein Ende? [No programmatic advertising without data is the GDPR now putting an end to automated advertising?]. https://www.absatzwirtschaft.de/ohne-daten-kein-programmatic-advertising-setzt-die-dsgvo-automatisierter-werbung-jetzt-ein-ende-220434. Accessed 22 June 2018
70. Bakare, S.O., Tolulope, O.: The unprotected port: a legal appraisal of DataDriven (programmatic) advertising in Nigeria. Int. J. Res. Innov. Social Sci. **05**(08), 525–532 (2021)
71. Alaimo, C., Kallinikos, J.: Objects, metrics and practices: an inquiry into the programmatic advertising ecosystem. In: Schultze, U., Aanestad, M., Mähring, M., Østerlund, C., Riemer, K. (eds.) Living with Monsters? Social Implications of Algorithmic Phenomena, Hybrid Agency, and the Performativity of Technology. IS&O 2018. IFIP Advances in Information and Communication Technology, pp. 110–123. Springer, Cham (2018)
72. Ciuchita, R., Gummerus, J.K., Holmlund, M., Linhart, E.L.: Programmatic advertising in online retailing: consumer perceptions and future avenues. J. Serv. Manag. **34**(2), 231–255 (2022)
73. Interactive Advertising Bureau: Programmatic and Automation – The Publishers' Perspective (2013). https://www.iab.com/wp-content/uploads/2015/06/IAB_Digital_Simplified_Programmatic_Sept_2013.pdf
74. Field, D., Zwillenberg, P., Rosenzweig, J., Zuckerman, N., Ruseler, M., Group, B.C.: The Programmatic Path to Profit for Publishers. https://www.bcg.com/publications/2015/media-entertainment-marketing-programmatic-path-profit-publishers. Accessed 17 July 2015

75. Sarna, N., Doligalski, T.: Programmatic advertising ecosystem as a multi-sided platform. An Evolutionary Perspective (2024)
76. Advance Metrics: Programmatic Advertising – Optimiertes Targeting [Programmatic advertising - optimised targeting]. https://www.advance-metrics.com/de/blog/programmatic-advertising-optimiertes-targeting/. Accessed 11 Nov 2021
77. Apps Flyer: Medieneinkauf auf Autopilot: Der ultimative Guide zu programmatic Advertising [Media buying on autopilot: The ultimate guide to programmatic advertising] (2024). https://www.appsflyer.com/de/resources/guides/programmatic-advertising/
78. Gonzalvez-Cabañas, J.C., Mochón, F.: Operating an advertising programmatic buying platform: a case study. Int. J. Interactive Multimedia Artif. Intell. **3**(6), 6–15 (2016)
79. Högström, J., Wallin, E.: Programmatic advertising: a quantitative study of consumer attitudes towards highly frequent online ads. [Bachelor thesis, Halmstad University] (2017)
80. Mahler, P.: Programmatic targeting: how to reach your audience. https://www.eskimi.com/blog/programmatic-targeting. Accessed 12 Jan 2024
81. Pawlata, H., Cakir, G.: The impact of the transparency consent framework on current programmatic advertising practices. In: Proceedings of the 4th International Conference on Computer-Human Interaction Research and Applications, vol. 1, pp. 331–336 (2020)
82. Senn, M.: Programmatic Advertising einfach erklärt Vom Anfänger bis zum Profi [Programmatic advertising explained simply from beginner to professional]. https://www.sidekicks.ch/programmatic-advertising-einfach-erklaert-vom-anfaenger-bis-zum-profi. Accesed 18 Jan 2023
83. Shehu, E., Abou Nabout, N., Clement, M.: The risk of programmatic advertising: effects of website quality on advertising effectiveness. Int. J. Res. Mark. **38**(3), 663–677 (2021)
84. WARC: What we know about reach and frequency (2025). https://www.warc.com/content/paywall/article/bestprac/what-we-know-about-reach-and-frequency/en-gb/107153
85. Advant Technology: How to calculate viewability for programmatic display & video ads. https://advanttechnology.com/how-to-calculate-viewability-for-programmatic-display-video-ads/
86. mso Digital: Viewability. https://www.mso-digital.de/wiki/viewability/
87. Ercan, H.D., Tanrıverdi, N.S., Taşkın, N.: A systematic literature review for artificial intelligence in advertising. In: Proceedings: International Conference on Information Resources Management (CONF-IRM), vol. 1 (2024)
88. AgencyAnalytics: Engagement Rate (2025). https://agencyanalytics.com/kpi-definitions/engagement-rate
89. Palla, P.J., Kyriacou, E., Zarkada, A.: The effect of involvement on attitude formation and strength in the digital domain. J. Mark. Commun. **29**(2), 137–160 (2022)
90. Zarkada, A.K., Panigyrakis, G.G., Tzoumaka, E.: Hosting a successful metamodern party: mixed methods management research on the web 2.0+. In: Moutinho, L., Sokele, M. (eds.) Innovative Research Methodologies in Management: Volume I: Philosophy, Measurement and Modelling, pp. 1–27. Springer, Cham (2018)
91. Zarkada, A.K.: Listening to a unicorn's pink narrative: Scheduling of brand Posts on Instagram for maximum consumer brand engagement. In: 19th International Conference on Research in Advertising (ICORIA) Virtual Conference (2021)
92. Chaffey, D.: What is programmatic marketing?. https://www.smartinsights.com/internet-advertising/internet-advertising-targeting/what-is-programmatic-marketing/. Accessed 22 Dec 2021
93. Chaffey, D., Ellis-Chadwick, F.: Digital Marketing: Strategy, Implementation and Practice. Pearson Education Limited (2022)
94. Davidova, E., Antuntes, A.C.: The impact of retargeting on consumer well-being and purchase intention. In: Remondes, J., Madeira, P., Alves, C. (eds.) Connecting With Consumers

Through Effective Personalization and Programmatic Advertising, pp. 203–220. IGI Global (2024)
95. AlSabeeh, D.A., Moghrabi, I.A.R.: Programmatic advertisement and real time bidding utilization. In: Kar, A.K., et al. (eds.) Digital Nations – Smart Cities, Innovation, and Sustainability, pp. 289–297. Springer, Cham (2017)
96. DataScientest: CPA (Cost per Acquisition): Was ist das? Wozu dient es? [CPA (Cost per Acquisition): What is it? What is it for?]. https://datascientest.com/de/cpa-cost-per-acquisition-was-ist-das-wozu-dient-es. Accessed 27 June 2023
97. Hübinette, S.: The true effects of programmatic display marketing: A study on how advertisers could make use of programmatic in the different stages of the customer journey [Master thesis, KTH Royal Institute of Technology] (2017)
98. SendPulse: Brand Recognition. https://sendpulse.com/support/glossary/brand-recognition. Accessed 28 Sept 2024
99. Tsai, Y.-L., Honka, E.: Informational and non-informational advertising content. SSRN Electron. J. (2020)
100. Hamelin, N., Al-Shihabi, S., Quach, S., Thaichon, P.: Forecasting advertisement effectiveness: neuroscience and data envelopment analysis. Australas. Mark. J. **30**(4), 313–330 (2021)
101. Wagener, A.: Marketing 4.0 [Digitalisation: blessing or curse: how digitalisation is changing the way we live and work]. In: Wolff, D., Göbel, R. (eds.) Digitalisierung: Segen oder Fluch: Wie die Digitalisierung unsere Lebens-und Arbeitswelt verändert, pp. 125–150. Springer Berlin (2018)
102. Adjust: What is return on ad spend (ROAS)? (2025). https://www.adjust.com/glossary/roas-definition/
103. Joosten, D.: Multi-touch attribution models: The ultimate guide. https://www.growthloop.com/university/article/multi-touch-attribution. Accessed 0 Nov 2024
104. Li, H., Kannan, P.K.: Attributing conversions in a multichannel online marketing environment: an empirical model and a field experiment. J. Mark. Res. **51**(1), 40–56 (2014)
105. Nielsen: Methoden und Modelle: Ein Leitfaden für Multi-Touch-Attribution [Methods and models: A guide to multi-touch attribution]. Nielsen (2019) https://www.nielsen.com/de/insights/2019/methods-models-a-guide-to-multi-touch-attribution/
106. Ohlinger, J.D., Nedyalkov, N.: Incrementality testing: the key to unlocking profitable growth in a changing industry. Think with google (2023); https://www.thinkwithgoogle.com/intl/en-emea/marketing-strategies/data-and-measurement/incrementality-testing/
107. Tharp, M.: The Pros and Cons of Incrementality Testing. https://www.rockerbox.com/blog/incrementality-testing-pros-cons. Accessed 20 Dec 2022
108. Lester, M., Charles, H.: Leveraging digitalization for SME growth: a roadmap to competitive advantage (2024)
109. Fernandez-Tapia, J.: An analytical solution to the budget-pacing problem in programmatic advertising. J. Inf. Optim. Sci. **40**(3), 603–614 (2019)
110. Chen, G., Xie, P., Dong, J., Wang, T.: Understanding programmatic creative: the role of AI. J. Advert. **48**(4), 347–355 (2019)
111. Klee, C.: AI in Marketing – Curse or Blessing? Impacts of Programmatic Advertising and Personalized Content on Society [Bachelor thesis, KTH Royal Institute of Technology] (2021)
112. Brandshark: Real-time bidding vs. Traditional advertising. https://brandshark.com/real-time-bidding-vs-traditional-advertising/. Accessed 23 June 2023
113. Google: Clickthrough rate (CTR): Definition (2025). https://support.google.com/google-ads/answer/2615875?hl=en
114. Hynes, E.: The Ultimate Programmatic Advertising Glossary. https://www.stackadapt.com/resources/blog/programmatic-advertising-glossary. Accessed 08 Nov 2022

115. Jabbari, C.: The ultimate programmatic advertising glossary (2025). https://www.outbrain.com/blog/programmatic-advertising-glossary/
116. Katana Media: Programmatic Buying 101: The Cost Efficiencies. https://medium.com/katana-media/programmatic-buying-101-the-cost-efficiencies-785b9a569171. Accessed 22 Mar 2017
117. Learn Ad Operations: The Role of Ad Exchanges and SSPs in Programmatic Advertising (2024). https://learnadoperations.com/the-role-of-ad-exchanges-and-ssps-in-programmatic-advertising/. Accessed 13 June 2024
118. Mailchimp: Conversion Rate: What Is It & How to Calculate It. (2024). https://mailchimp.com/marketing-glossary/conversion-rates/
119. Muthukrishnan, S.: Ad exchanges: research issues in internet and network economics. In: Leonardi, S. (ed.) WINE 2009. Lecture Notes in Computer Science. Springer, Heidelberg (2009)
120. smartclip: Reach. https://smartclip.tv/adtech-glossary/reach/. Accessed 15 Mar 2022
121. Hur, J.: The History of Paid Search Engine Marketing (2024). https://bebusinessed.com/history/the-history-of-paid-search-engine-marketing/

Author Index

A. K. Zarkada (Ed.): GMA-GAMMA 2025, SPBE, p. 241, 2025.
https://doi.org/10.1007/978-3-031-96930-0